T0181757

Advances in Geographic Information Science

Series editors

Shivanand Balram, Burnaby, Canada
Suzana Dragicevic, Burnaby, Canada

More information about this series at http://www.springer.com/series/7712

Ola Ahlqvist • Christoph Schlieder

Editors

Geogames and Geoplay

Game-based Approaches to the Analysis of Geo-Information

 Springer

Editors
Ola Ahlqvist
Department of Geography
The Ohio State University
Columbus, OH, USA

Christoph Schlieder
Faculty of Information Systems
and Applied Computer Sciences
University of Bamberg
Bamberg, Germany

ISSN 1867-2434   ISSN 1867-2442 (electronic)
Advances in Geographic Information Science
ISBN 978-3-319-79423-5   ISBN 978-3-319-22774-0 (eBook)
https://doi.org/10.1007/978-3-319-22774-0

Printed on acid-free paper

This Springer imprint is published by Springer Nature
The registered company is Springer International Publishing AG
The registered company address is: Gewerbestrasse 11, 6330 Cham, Switzerland

Foreword

Many people around the world were taken by surprise in July 2016 by the release of Pokémon Go, a virtual-reality-based mobile game. The mainstream press felt the need to expose every possible angle to the story: privacy, safety, physical exercise, add-on complementary products, and of course business models.

But this mobile game did not appear from a vacuum. It was a re-skinning (rebranding) of a far less popular game called Ingress, which served for a couple of years as the user-generated data collector that fed the current version. The people behind the game at the company Niantic had not only top pedigree (the CEO John Hanke was a co-founder of Keyhole, which became Google Earth) but also the backing of the technology giant Google/Alphabet from which it spun-off. And the partnership with Pokémon's creator, Nintendo, did not hurt either.

This is one of the relatively few success stories in the world of thousands of game launches every year. The editors and authors of this book have been studying similar games—geogames—and related game mechanics for years now. Successful geogames are born of successful planning, narrative design, technical implementation, marketing, as well as other factors. These can and are being studied in much greater depth at universities around the world.

When I contacted the editors to ask their opinion on the new Pokémon Go, they took some time to study it deeply and came back with some interesting, detailed criticism (and praise). Just as a best-selling novel might not be the best-written, this hugely popular game had its flaws. Some—excessive personal data collection for example—were picked up by others, and Niantic was forced to make immediate fixes. But the point is that geogames—mobile games in which geographic location of the players is a foundational characteristic—are easier than ever to create; however, creating a *successful* geogame remains as much an art as a science.

This book covers many of the key aspects of geogame and geoplay design, implementation, and testing. Of special interest to me are two concepts: geogame patterns and relocation from one context (city) to another. Identifying patterns allows us to more easily abstract and to imagine how new game ideas can fit into an overall structure and therefore borrow or inherit well-tested ideas from nearby fields or communities. Relocation of a geogame from city to city involves interesting geographic information system (GIS) tasks such as identifying similar places by their geometrical or descriptive characteristics.

It has been a pleasure to have worked with the editors and with some of the authors over the last few years, in the ideation, testing, and creation of geogames. This book also is the fruit of several workshops on geogames and geoplay, held in locations such as California, Spain, Austria, and Finland, during which very useful feedback was received from a wide range of participants, from programmers to educational psychologists.

I hope that the reader of this book also provides feedback and actively participates in this nascent community of geogame and geoplay researchers and practitioners. This community will surely grow and prosper in the coming decades and will be able to point to this book as an early anchor or flag planted in the sand. Imagine, create, explore, learn, enjoy.

Michael Gould
Esri, Inc.
Redlands, CA, USA
University Jaume I
Castellón, Spain

Contents

Contributors

Ola Ahlqvist Department of Geography, The Ohio State University, Columbus, OH, USA

Vyron Antoniou Hellenic Army General Staff/Geographic Directorate, Athens, Greece

Emelie Bailey The Ohio State University, Columbus, OH, USA

Thomas Bartoschek Institute for Geoinformatics, University of Muenster, Muenster, Germany

Rohan Benkar The Ohio State University, Columbus, OH, USA

Vânia Carlos University of Aveiro, Aveiro, Portugal

Michael N. DeMers New Mexico State University, Las Cruces, NM, USA

Andrew Heckler Ohio State University, Columbus, OH, USA

Thomas Heinz University of Bamberg, Bamberg, Germany

Nathaniel J. Henry Ohio State University, Columbus, OH, USA

Christopher Holden University of New Mexico, Albuquerque, NM, USA

Swaroop Joshi Ohio State University, Columbus, OH, USA

Peter Kiefer Institute of Cartography and Geoinformation, ETH Zurich, Zurich, Switzerland

Dominik Kremer University of Bamberg, Bamberg, Germany

Rui Li University at Albany-SUNY, Albany, NY, USA

Stefan Münzer University of Mannheim, Mannheim, Germany

Jim Mathews Field Day Lab, University of Wisconsin, Madison, WI, USA

Alenka Poplin Iowa State University, Ames, IA, USA

Rajiv Ramnath Ohio State University, Columbus, OH, USA

Simon Scheider Institute of Cartography and Geoinformation, ETH Zurich, Zurich, Switzerland

Department of Human Geography and Spatial Planning, University Utrecht, Utrecht, Netherlands

Christoph Schlieder University of Bamberg, Bamberg, Germany

Angela Schwering Institute for Geoinformatics, University of Muenster, Muenster, Germany

Neelam Soundarajan Ohio State University, Columbus, OH, USA

Kiril Vatev Ohio State University, Columbus, OH, USA

Kavita Vemuri International Institute of Information Technology, Hyderabad, India

Cheng Zhang NASA Scientific Visualization Studio at Goddard Space Flight Center, Greenbelt, MD, USA

The Ohio State University, Columbus, OH, USA

Chapter 1
Introducing Geogames and Geoplay: Characterizing an Emerging Research Field

Ola Ahlqvist and Christoph Schlieder

1.1 Introduction

Games and play are part of human life, and place, space, and geography take central roles in determining the rules and interactions of games. Consider how integral maps are to the board game RISK, how video game players navigate through a realistic 'world' in pursuit of a goal, the millions of Pokemon Go players navigating the real world to find new Pokemon. Even the very abstract maps of Monopoly and Chess are inherently geographical, utilizing basic spatial rules for game play.

This is a book about games that use real-world information and geographic information technologies. As such it is the first of its kind, which may sound surprising considering the ubiquity of geographic information (GI) technologies and location-based applications in modern life. We use navigation apps to find our way from A to B, we check in at places with social apps, and digital photographs typically contain the location of each picture as part of the data file. However, this widespread access to mobile information, communication, and geospatial technologies only very recently inspired any significant development of gaming activities that are connected to the real world. Terms for these games began appearing around 2004–2005: pervasive games, ubiquitous games, augmented-, alternate-, and mixed-reality games, mobile games, and adaptronic games.

We seek to provide a wide umbrella for this first book, and with the title *Geogames and Geoplay* we consider all games and play that uses real geocontent and is

O. Ahlqvist (✉)
Department of Geography, The Ohio State University, 1049B Derby Hall, 154 N Oval Mall, Columbus, OH 43210, USA
e-mail: ahlqvist.1@osu.edu

C. Schlieder
Faculty of Information Systems and Applied Computer Sciences, University of Bamberg, Kapuzinerstraße 16, 96047 Bamberg, Germany

1

mediated by GI technology. In the following subsections we will elaborate further on this definition, and the chapter concludes with an introduction to the rest of this book. The authors of the book chapters are representative of a diverse community of researchers and designers, a community, we must add, which has not agreed on a standard terminology for describing geographic gameplay so far. We have refrained from asking the authors to commit to a list of technical terms created by the editors because we could foresee that they would return to their preferred terminology in their very next publication. We have taken care, however, that each chapter defines its core concepts. A prominent case is the term "geogame" itself, which some chapters use in a narrower sense than we do here.

1.2   GIS/Spatial Principles and Game Patterns

A key motivation for this book was the conception that geocontent, i.e. real-world spatial information, brings something unique to games and play. Many of the terms mentioned above (mobile, location-based, etc.) have become associated with particular technologies, which means that they would be subject to change with constantly changing technology. We therefore seek a more stable and generic way to characterize and define the realm of "Geogames and Geoplay".

This calls for an examination of what elements define the geographic dimension and the gaming dimension of geogames. In the following sections we will interrogate some of the existing literature on the core concepts for geographic information and game design. After this, we will provide two different perspectives on how the two realms of Geo and Games map onto each other.

1.2.1   Core Geographic Concepts

In an effort intended to support "…a broader use of spatial information in science and society", Kuhn (2012) sought to provide an understanding of what spatial information is and how it can be used across disciplines and populations in science and society. He identified ten core concepts of *spatial information* to guide experts, scientists and practitioners across a wide range of disciplines. The ten concepts are: Location, Neighbourhood, Field, Object, Network, Event, Granularity, Accuracy, Meaning, and Value. Although Kuhn and Ballatore (2015) later narrowed this list of core concepts down to seven, eliminating Neighborhood, Meaning, and Value, because we are interested in identifying a comprehensive collection of core concepts, we will consider the original ten in our following discussion.

Another theory of core geographic concepts related to *spatial thinking* was developed around the same time, but independently of Kuhn, by Janelle and Goodchild (2011). Their concepts were a synthesis of several previous works and identified the following nine as a foundation for *spatial reasoning*: Location,

Distance, Neighborhood and Region, Networks, Overlays, Scale, Spatial Heterogeneity, Spatial Dependence, Objects and Fields. It should be noted that these two proposals are slightly different in their objectives. While Kuhn's focus is on spatial information and its properties, Janelle and Goodchild focus on spatial thinking in the social sciences. So, by examining the two proposals side by side (Table 1.1) we find significant overlaps, but also some unique contributions that forms a tentative synthesis of the two perspectives.

Kuhn does not identify distance per se as a key concept, but rather suggests that "nearness" is "...a natural companion concept to location" and part of his "Neighbourhood" concept. Other than those small nuances, the first halves of the two frameworks, as listed above, largely agree (see Table 1.1).

Differences are most clear upon reaching Kuhn's (2012) concept of Event (time), and the non-spatial information concepts of Accuracy, Meaning and Value. Janelle and Goodchild include the analytical concepts of Overlays, Spatial Heterogeneity and Spatial Dependence. It seems clear from Janelle and Goodchild's presentation that time is considered an integral component of their Fields/Objects concept where they explicitly talk about space-time. The remaining differences are clearly related to the different perspectives taken by each proposition. Accuracy, Meaning and Value are all specific and central to geographic *information*, whereas Overlays, Spatial heterogeneity and Spatial dependence are all foundational ideas to spatial *thinking*.

Thus, we will consider the unified list (right column in Table 1.1) as a preliminary comprehensive list of core concepts for geographic information and thinking. With this conceptual framework in place for the "Geo" realm, we now turn our attention to the "Game" realm.

1.2.2 Core Game and Play Concepts

In game design there is no direct equivalent to the core geographic concepts. Game designers have looked, however, for methods and typologies assisting them in reducing the complexity of their task. As in other design disciplines, most notably in architecture and software engineering, a compositional approach has proven successful. This approach describes a design as consisting of interrelated conceptual building blocks, so-called patterns. When Björk, Lundgren, and Holopainen (2003) introduced patterns into game research, they did so by explicitly referring to the already well-established software design patterns of Gamma et al. (1995). Using that perspective, a game pattern describes a generic solution to a specific class of design problems. The high-score list pattern, for instance, solves the problem of instigating competition among players without implementing a full-blown multi-player game flow. Like in software engineering, patterns are associated with trade-offs and design implications, which are included in the pattern description.

Table 1.1 Alignment and differences between core geographic concepts proposed for spatial information (Kuhn 2012; Kuhn and Ballatore 2015) and for spatial thinking (Janelle and Goodchild 2011)

Kuhn (2012), Kuhn and Ballatore (2015)	Janelle and Goodchild (2011)	Proposed unified core geographic concepts
Location—answers *where?* but should be understood as a relation between figures located with respect to a chosen ground	*Location*—formal and informal methods of specifying "where" (locations and divisions of the world)	**Location**—formal and informal ways of specifying *where?*
	Distance—relationships between places by measures of proximity	**Distance**—relationships between places/objects by measures of (space-time) proximity
Neighbourhood—answers *what is near?* commonly thought of as regions	*Neighborhood and Region*—drawing inferences from spatial context (situations and neighborhood of places)	**Neighborhood**—identifies places/regions in terms of distance and spatial context
Field—describe continuous phenomena that have an attribute everywhere in a space of interest	*Fields*[a]—phenomena that are continuous in space-time	**Field**—continuous phenomena that have a thematic attribute everywhere in space-time
Object—describe individuals or elements that have an identity as well as spatial, temporal, and thematic properties	*Objects*[a]—phenomena that are discrete in space-time	**Object**—individuals that have an identity as well as spatial, temporal, and thematic properties
Network—connectivity as captured by binary relationships between nodes (objects)	*Networks*—linear networks with connections and flows	**Network**—relationships (connections and flows) between objects (nodes)
Granularity—amount of detail in spatial information	*Scale*—level of detail in a geographic dataset	**Granularity/Scale**—extent and amount of detail in spatial information
Event—an individual portion of a process, bounded in time		**Event**—an individual portion of a process, bounded in time
	Overlays—inferring spatial associations by comparing mapped variables by locations	*Overlays*—inferring spatial associations by comparing mapped variables by locations
	Spatial heterogeneity—the implications of spatial variability	*Spatial heterogeneity*—implications of spatial variability
	Spatial dependence—understanding relationships across space	*Spatial dependence*—understanding relationships across space

(continued)

Table 1.1 (continued)

Kuhn (2012), Kuhn and Ballatore (2015)	Janelle and Goodchild (2011)	Proposed unified core geographic concepts
Accuracy—difference between spatial information and some reference considered 'true'		**Accuracy**—difference between spatial information and some reference considered 'true'
Meaning—how to interpret terms in spatial information		**Meaning**—how to interpret terms in spatial information
Value—the roles played by spatial information in society		**Value**—the roles played by spatial information in society

[a]Objects and Fields are presented together as one concept by Janelle and Goodchild

Two inventories of game patterns are of special interest to the study of geogames and geoplay. Davidsson et al. (2004) compiled a collection of 74 patterns. It is less comprehensive than the list identified by Björk and Holopainen (2004), but in our context, it has two advantages: First, the patterns are specific to games played on mobile devices, while the more extensive list refers to all kinds of games from pre-computer children's puzzles to survival horror video games. Second, and more important, the identification of the pattern inventory relies on a systematic method in which expert game designers analyze a set of well-known model games. More recently, Sintoris (2015) has identified another inventory of 41 game patterns using a comparable method and focusing on location-based games only (all studied games use some kind of positioning technology and let the players interact with the geographic environment). Detailed pattern descriptions are available on the project's Wiki (Sintoris 2015).

In the context of geogames that are not location-based, it is useful to also look at inventories of general, not just spatial, game patterns. Building on earlier works on game patterns, Holopainen (2011) developed a component framework to help navigate and define general design patterns that have been identified. He identified 18 components as the "basic building blocks" of games and grouped them into four categories: (1) Holistic—These components help in defining how gaming differs from other activities and describing how players can join and end a specific game. The holistic components are: **Game Instance, Game Session, Play Session, Set-up and set-down sessions**, and **Extra-Game Activities** that are related to but not directly affecting game play itself. (2) Bounding—These components define purposes for playing the game and permissible game play activities. The bounding components are: **Rules, Modes of Play, Goals and Sub-goals**. (3) Temporal—These components record the game play by identifying and separating a larger game play activity into temporally separated activities. The temporal components are: **Actions, Events, Closures, End Conditions, Evaluation Functions**. (4) Structural—These components describe the basic parts of the game, such as objects representing real-world or imaginary objects, people or creatures, or abstract phenomena like values or attributes. The structural components are: **Game Facilitator,**

Players, Interface, Game Elements, Game Time—In a similar effort, and also informed by the early work on game design patterns, Järvinen (2008) identified nine Key Game Elements as generic classes of things that make up an entire game system and suggested they could be used for analyzing games. They are: **Components, Rule Set, Environment, Game Mechanics, Theme, Information, Interface, Player(s)**, and **Contexts**. While Holopainen's work primarily uses a game design perspective, the elements identified by Järvinen is derived from a more user oriented perspective. Maybe due to this difference the correspondence of these two typologies are not as apparent as with the previous spatial concepts, but examining them side by side does allow us to assess possible alignments as well as unique ideas, just as we did with the spatial concepts in Table 1.1.

As we can see in Table 1.2, these two separate efforts display some similarities and overlaps, but also significant differences. Maybe the largest difference is Holopainen's (2011) temporal components of Game instance, Game session, Play session, Set-up & set-down, Extra game activities, and Game time, which do not have a direct equivalent in Järvinen's (2008) game elements. Järvinen discusses these in terms of 'game states' as temporal reference points, but never raises it to the level of a separate game element. Rather, it is something that his compound 'Information' element would capture through query of the system. The benefits of Holopainen's more specific set of concepts is that we get a vocabulary for talking about a simulation as a closed event with particular sub-components. Similarly, Holopainen provides a more specific vocabulary for the rules, policies, goals and events that emanate from play activities. These are all more or less subsumed by Järvinen's much broader term "Rule set", but they identify a collection of unique temporal concepts related to specific aspects of game events and procedures that are ultimately governed by rules. Järvinen's game elements Contexts and Theme do not have a direct correspondence in Holopainen's components. This is probably due to Järvinen's focus on semiotic, rhetorical and cultural aspects of games. As we will see in the next section though, these elements seem to have direct alignments with some of the core geographic concepts identified previously. For the remaining elements and components there seems to be a fairly obvious alignment of concepts related to interface, player actions, and the game 'world'.

1.3 Reconciling Core Geographic and Game Concepts

We are now ready to examine the geographic and game dimensions together, as articulated in the core concepts identified in the previous section. We will first highlight how spatial information processing relates to the game patterns from the inventory of Davidsson et al. (2004) and the inventory of Sintoris (2015), we then provide a complementary perspective on how the core concepts of games and play from Table 1.2 can enrich the geographic core concepts in Table 1.1.

Table 1.2 Alignment and differences between proposed game components (Holopainen 2011) and key game elements (Järvinen 2008)

Game components	Key game elements
Holopainen (2011)	Järvinen (2008)
Game Instance—a single completion of game play as a unique configuration of a set of players, the place where it is played, and external circumstances under which it is played **Game Session**—the activity undertaken in a game instance by the game's players **Play Session**—several distinct periods of game play activity inside a game session **Set-up and Set-down Sessions**—game and play activities that do not constitute game play directly but are required and take place nevertheless **Extra-Game Activities**—any extra activities related to the game but not directly related to playing the game itself. **Game Time**—the timeline of sequentially ordered actions in a game session	**Information**—information about events, agents, objects, and the system state
	Contexts—where, when and why the game encounter takes place.
Rules—explicit or implicit policies that dictate the flow of the game **Modes of play**—sections, phases or turns where the interface, available actions, and information for the players—and thus also the activities—are very different. **Goals and sub-goals**—the aim of players' plans and actions in a game **Events**—discrete points in game play where the game state changes **Evaluation Functions**—determines the outcome of an event **Closures**—the completion of a goal or a sub-goal **End conditions**—the game states when closures occur and when the game instance ends.	**Rule set**—procedures that governs game play, permissible actions, etc.
Actions—discrete or continuous player actions that change the game state	**Game mechanics**—actions that players can engage in as part of playing, e.g. placing, shooting, maneuvering, trading.
	Theme—the subject matter of a game, often used to provide a metaphor for the game system and rule set, e.g. a treasure hunt or a command and conquer historic war scenario.

(continued)

Table 1.2 (continued)

Game components	Key game elements
Holopainen (2011)	Järvinen (2008)
Game Facilitator—oversees the workings of a game, taking care of the game events, rules and also resolves possible disputes. **Players**—the entities that strive toward the goals in a game by choosing and performing actions.	**Players**—people who play
Interface—the different types and forms of representations by which players have access to a game	**Interface**—various means, devices and tools that allow players to access and interact with the other game elements, e.g. pen and paper, computer screen, pointing device.
Game Elements—the physical and logical attributes that help maintain and inform players about the current game state	**Components**—the resources for play that are moved or modified in game transactions, e.g. tokens, tiles, characters, vehicles. **Environment**—the space for play, e.g. a specific setup, game board, grid, maze, level, world.

1.3.1 Spatial/Non-spatial Game Patterns and Core Geographic Concepts

While the game pattern inventories (Davidsson et al. 2004; Sintoris 2015) described in the previous section do not cover the entire field of geogames and geoplay, they nevertheless provide us with an opportunity to assess the role of spatial concepts in an important subfield of spatial game design. We evaluated the pattern descriptions of the two inventories using the core spatial concepts (Table 1.1) occurring in them. Verbatim occurrences (e.g. "proximity") as well as paraphrases (e.g. "the distance between the player and a certain physical location" for "proximity") were noted, and patterns, which explicitly referred to at least one core concept were categorized as having a strong spatial component. The analysis of the pattern descriptions from Davidsson et al. (2004), reveals that 23 (31%) of the 74 patterns have such a strong spatial component. The "Strategic Locations" pattern is one example. Its description states: "Mobile games … often make use of strategic locations where players receive special benefits". A portal in the Ingress game, for instance, constitutes such a strategic location that players must interact with in order to win the game. Most patterns, however, are non-spatial like the "Highscore List" pattern. Interestingly, the analysis of the inventory of Sintoris (2015) produces a similar result.

Again, a part of the patterns, 13 (32%) from 41, show a strong connection to spatial information. Considering both inventories, we find that, roughly speaking, one third of the game patterns are spatial. To a certain extent, this explains why a designer may get along for some time without caring much about spatial information processing. Location-based geogames are first and foremost games. Spatial information processing is secondary and only involved to the extent needed to support

Table 1.3 Fundamental spatial game patterns

	Characteristics of the cluster of game patterns	D = Davidsson et al. (2004) S = Sintoris (2015)	Associated core concepts
Locality	entering (or leaving) a place enables specific game actions	Strategic locations (D 28)	Location object
		Spatial structure (S 6.4)	
		Co-locality (S 3.18)	
Proximity	approaching (or gaining distance from) a player or artifact triggers game events	Player-location proximity (D 8)	Distance neighborhood field overlays
		Artifact-location proximity (D 9)	
		Player-player proximity (D 10)	
		Artifact-artifact proximity (D 13)	
		Proximity (S 6.2)	

game design goals. However, with one-third of the patterns being related to spatial concepts, these patterns bear considerable potential to improve a design. This is where the geographic core concepts can demonstrate their utility. They allow us to establish a correspondence between spatial game patterns and the core concepts, which reveals in which respect both inventories are still incomplete. Note that the two inventories (Davidsson et al. 2004; Sintoris 2015) do not identify exactly the same set of spatial patterns. This is not too surprising as they were elicited from two different sets of model games. Since we are interested in a common core of patterns, we used the underlying concepts of spatial information processing to group the game patterns from both inventories into clusters. The grouping is based on the pattern descriptions found in the two inventories. We group together patterns, which predominantly refer to the same core concept.

At the most basic level, two clusters of patterns emerge: locality and proximity (Table 1.3). Locality patterns refer to places, that is, to geographic positions with a game-specific meaning (e.g. a portal in the Ingress game). Most game designers model such places as objects with spatial boundaries. The spatial position of the player with respect to the places determines the available game actions. Locality patterns require some positioning technology that determines automatically whether the player enters a place and may perform the intended action. Proximity refers to the concepts of distance and neighborhood, and the variation of proximity constitutes a (local) field where the closeness to a location, to another player, to a virtual or a real artifact triggers a game event. Such patterns are based on a gradual decision function, which is why the inventory of Sintoris (2015) associates these patterns with the hot-cold type of feedback given to players.

Table 1.4 Spatial activity game patterns

	Characteristics of the cluster of game patterns	D = Davidsson et al. (2004) S = Sintoris (2015)	Associated core concepts
Navigation	identifying and following a route in geographic space	Physical navigation (D 5)	Network granularity
		Path-finding (S 3.7)	
Race	Reaching a place before other players do	Race (D 31)	Event accuracy
		Competition (S 1.3)	
Collection	Identify, locate and get hold of an object	Collection (D 71)	Event value
		Collecting (S 3.17)	

The locality and proximity patterns are fundamental in the sense that any location-based game engine has to support at least one such pattern. Other game patterns refer to more complex core concepts like network, granularity, event, accuracy or value. It is possible to identify associated clusters of patterns describing different spatial activities (Table 1.4). We found three such clusters in the two inventories: navigation patterns, race patterns, and collection patterns. Interestingly, some patterns frequently found in location-based games are missing. Two examples are spatial exploration patterns such as hierarchical spatial search and spatial movement patterns such as flock movement or encirclement. It would be extremely helpful for the design of spatial game engines to have a complete list of spatial activity patterns. Anyone attempting to do a complete review of spatial activity patterns should start with a more comprehensive list of model games. We must leave this task to future research.

Besides the fundamental spatial patterns (Table 1.3) and the spatial activity patterns (Table 1.4) there is a third group of patterns, which we might call spatial interaction patterns. They are more complex in that they involve social conventions and interactions. We will not list them in a table because the inventories mention only a single one: the gaining ownership pattern (No. 29 in Sintoris 2015). The dearth of spatial interaction patterns suggests that the inventories are quite incomplete.

Although our observations on the importance of the core concepts to game design are restricted to location-based geogames, they may transfer to other geogames, for instance, geographic simulation games. Very likely, an analysis of a set of model simulation games will also reveal the fundamental patterns of locality and proximity. The spatial activity patterns (e.g. race) might be different and the social interaction patterns (e.g. ownership) will most likely be more complex in simulation games than in location-based geogames.

1.3.2   Intersection of Core Geographic Concepts with Games and Play

There are also some direct correspondence or similarity between the game components/elements (Holopainen 2011; Järvinen 2008) and core geographic concepts. In an effort to identify those overlaps and to possibly enrich the core geographic concepts with some of the unique aspects that games embody, we have tried to align the core geographic concepts from Table 1.1 with the two described typologies for game analysis from Table 1.2. Our proposed alignment can be found in Table 1.5 below, where we list the previously proposed collection of unified core geographic concepts in the left column and the game components and elements in the middle and right column. In addition, we tentatively suggest Rules, Agents, Interface, and Simulation as new concepts to be added to the set of core geographic concepts in order for them to capture the particulars of not only geogames but geographic simulations in general. These additional concepts will also be discussed below.

Some of Järvinen's elements have close correspondence with the core geographic concepts: **Components** and **Environment**—roughly corresponding to the discrete objects and continuous fields (or tessellations) of spatial information; **Theme**—roughly corresponding to that of thematic maps, determining the subject matter and its meaning.

A few other elements closely correspond to other well-known geographic concepts, yet these are not listed among the 'core' concepts: **Ruleset**—roughly corresponding to rules in spatial simulation and modelling; **Interface**—corresponds with the devices used to interact with the GIS, usually a computer or mobile device; **Players**—roughly corresponds to GIS users with the important caveat that it does not include game designers, hence not including for example GIS programmers, database managers, data producers etc. Although, in a spatial simulation context, players would more likely be equated with agents as in Agent Based Modelling, and Holopainen's (2011) game components related to game and play sessions closely correspond to the act of running a **Simulation** of a geographic system model. The fact that these are not directly associated with any of the Kuhn and Ballatore/Janelle and Goodchild core concepts suggests that the area of geogames offers an opportunity for identifying important gaps in the set of core geographic concepts. We propose variants of these existing concepts from the game design literature as four new additions, **Rules**, **Agents**, **Interface** and **Simulation**, to enrich the key concepts of geographic information and thinking.

A cross-cutting element in Järvinen's (2008) collection is **Information**, indicated by (I) in Table 1.4. This element pertains to all information about objects, events, agents, and the game system that is needed for the game to work. As such, it roughly corresponds to everything in a Geographic *Information* System that pertains to a particular (thematic) project, including its actors, available information, and constraints. Information is a key ingredient of the core geographic concepts and could therefore potentially help to further specify and organize Järvinen's Information element. Together with the suggestions in the previous section, this

Table 1.5 Alignment and differences between core geographic concepts and proposals for key game components (Holopainen 2011) and game elements (Järvinen 2008)

Proposed unified and *extended* core geographic concepts	Game components Holopainen (2011)	Key game elements Järvinen (2008)
Location—formal and informal ways of specifying *where?*		(I)
Distance—relationships between places/ objects by measures of (space-time) proximity		(I)
Neighborhood—identifies *what is near?* in terms distance		(I)
Field—continuous phenomena that have a thematic attribute everywhere in space-time	Game elements	Environment (I)
Object—describe individuals that have an identity as well as spatial, temporal, and thematic properties	Game elements	Components (I)
Network—relationships (connections and flows) between objects (nodes)	Game elements	Environment (I)
Granularity/Scale—amount of detail in spatial information		(I)
Event—an individual portion of a process, bounded in time	Actions (Events)	Game mechanics (I)
Overlays—Inferring spatial associations by comparing mapped variables by locations		(I)
Spatial heterogeneity—implications of spatial variability		(I)
Spatial dependence—Understanding relationships across space		(I)
Accuracy—difference between spatial information and some reference considered 'true'		(I)
Meaning—how to interpret terms in spatial information		Theme
Value—the roles played by spatial information in society		Contexts
New—**Rules**—Procedures that dictate and governs game play (simulation), permissible actions, etc.	Rules Modes of play Goals and sub-goals (Events) Evaluation functions Closures End conditions	Rule set
New—**Agents**—Entities that act toward the goals or just following rules in a game.	Game facilitator Players	Players (I) (Components) (Environment)
New—**Interface**—the different types and forms of representations by which players have access to a game	Interface	Interface

(continued)

Table 1.5 (continued)

Proposed unified and *extended* core geographic concepts	Game components Holopainen (2011)	Key game elements Järvinen (2008)
New—Simulation—dynamic enactment of interrelated objects, agents, and rules in a geographic system model	Game instance Game session Play session Set-up and set-down Extra game activities Game time	(I)

points at the potential for an enrichment of game design vocabularies using core geographic concepts.

The correspondence with the remaining two game elements to GIScience concepts is less clear. One is Järvinen's central concept of **Game Mechanics** that refers to the means by which players interact with game elements to influence game states in order to complete a goal. The 44 game mechanics that Järvinen (2008) identifies are divided into six categories, largely following the previous seven key elements: Component, Environment, Theme, Interface, Physical, and Player mechanics. The reason for this is that the nature of game mechanics is analogous with verbs such as "Arranging", "Trading", and "Moving", and the other elements mainly relate to the subject matter of the game, e.g. the components, environment and players that are involved in doing something. An example parallel to game mechanics in the context of GIS would be a spatial decision making or planning scenario, where one or more stakeholders seek to achieve some identified goal, e.g. optimal location of a new development, by using GIS information, analysis and modelling. An important difference is the richness of detailed mechanics identified for games because of their focus on player actions and interactions with any and all game components and the environment. We find Game Mechanics to most closely align with Kuhn's (2012) Event and the proposed 'new' concept of simulation. Much like the discussion of spatial activity patterns (Table 1.3) revealed, this is an area where much research remains to be done.

The second of Järvinen's (Ibid.) game elements that is difficult to match up with existing GIS concepts is **Contexts**, which refers to the time and place a game is played. This may sound like a repeat of the components and environment of the game, but it relates more broadly to a range of factors and relations to audience, cultures, traditions, public opinions, and motivations surrounding the game play. As such, game context seeks to consider social and psychological aspects of game play, and this seems most closely associated with the GIScience literature on cognitive (Nyerges et al. 1995) and societal (Nyerges et al. 2011) aspects of GIS practice. We find this to be most closely corresponding to Kuhn's (2012) Value concept, and this is another area of increasing importance where there is a lack of research.

Overall, Table 1.4 has helped us identifying the intersection of games and play with the geographic dimension. It illustrates that the core geographic concepts of location and analysis of space-time distance, neighborhood, overlap, heterogeneity, and dependence provide a specific and well-established vocabulary for defining

Geogames and Geoplay as a particular form of games and play that emphasize spatial relationships and patterns. Additionally, the game research vocabularies we have introduced here provide a rich source to draw from as we seek to develop a vocabulary to further describe Geogames, Geoplay, and geographic simulation activities. As we pointed out before, the notion of Rule Sets, and Players have close correspondence with spatial simulation concepts.

A significant body of GIScience research have pointed at the importance of interface modalities and it seems likely that this has a particularly important place in defining Geogames and Geoplay. In games it is quite common that the interface defines some of the game mechanics and ultimately the game itself (e.g. tennis, pinball, card games). To a certain degree, much recent development of mobile, location-aware devices have paved the way for many new Geogame ideas. Finally, Holopainen's (2011) temporal components of Game session, Play Session, Set-up and set-down, and Extra game activities have direct relevance to the idea of a spatial simulation. The benefits of Holopainen's more specific set of concepts is that we get a vocabulary for talking about a simulation as a closed event with particular sub-components.

The simulation aspect of Geogames and Geoplay is probably one of its most distinctive features. Järvinen (2008) argues that games should be seen as systems, including the dynamic interaction of objects, agents and events. This forms a foundation for using Geogames as instantiations of real-world systems, abstracted to some thematic focus, and able to provide insights into that system's behavior under the particular context of the game play.

We hope that this overview, even if it only constitutes a first effort, provides a more unified vocabulary that cuts across the realms of Geo and Games. As such it may provide a helpful foundation for reading this book and potentially for further research in this exciting area.

1.4 Structure of the Book and Research Questions

The purpose of this book is to provide a first overview of this highly interdisciplinary field with contributions from researchers, GIS professionals and game designers. Over the past 5 years we have seen a significant increase in the number of initiatives and efforts to build, disseminate and interrogate geogames and geoplay. There is an emerging research community sharing developments and insights in geogames and geoplay: A number of dedicated workshops have been held at international conferences, and funding agencies are now responsive to the emerging opportunities for discovery and societal impacts offered by this field. Slowly, this pioneering work is also beginning to provide the contours of a more comprehensive research agenda.

The chapters in this book are fairly representative of ongoing work as it cuts across several application areas, types of geogame and geoplay activities, and varying types of research approaches and traditions. The scope ranges from fundamentals about games and play, geographic information technologies, game design and culture, to current examples and forward looking analysis. Unlike other publications in this area (e.g. Nijholt 2017), where many of the represented perspectives come from planning, architecture, game studies, computer graphics, the perspectives provided in this book come primarily from the geospatial sciences. It therefore serves as an introduction for both geospatial scientists who seek orientation in the spatial gaming field, as well as for the aforementioned disciplines to gain an understanding of how those working in the spatial sciences may approach spatial games. Throughout the chapters, the authors refer to a number of different games. To facilitate citation and access to information on these games, we have compiled a joint ludography. In addition to games, the ludography also lists platforms and frameworks used for creating games. Games and platforms that have an entry in the ludography are identified by bold face. Classical console games are included even if they do not involve geographic gameplay. Traditional board and card games, however, are omitted. The ludography provides pointers to the chapters and, where possible, references to publications or links to websites.

In Chap. 2, Ola Ahlqvist, Swaroop Joshi and colleagues identify and describe the defining characteristics of a recently developed Geogame concept, a Geographic Information System-Multiplayer Online Game framework (GIS-MOG). This turns digital world maps, similar to Google and Bing Maps, into a game board where any place in the world can be experienced first-hand through board game-like simulations. The authors seek to define what it is about this technology that makes it a unique genre of geogames and learning technologies in general. Using the core concepts discussed in this introductory chapter, they provide a detailed description of how these concepts are incorporated in their geogame technology giving examples from their own game.

Chapter 3 by Thomas Bartoschek, Angela Schwering, and colleagues addresses the challenges of designing an educational game, **OriGami**, with the goal of improving the spatial orientation skills of young players in the age group of 8–12 year olds. The game can be played on mobile devices in outdoor environments, however, the chapter concentrates on the desktop mode. Their basic idea consists of letting the players solve route-following tasks based on verbal descriptions that mobilize different cognitive skills. An instruction such as "Turn right" may force the player to realign his or her cognitive map with the cartographic representation shown on the screen. The authors take an approach that is special in that it concentrates on a single skill, route following, and carefully analyses the task requirements. Even those who do not go through the details of the empirical study, will realize that the authors address a fundamental issue for the design of any educational game: Is there anything to learn? It is well-known that there are individual differences in spatial abilities and a game task has to be designed in a way that players can improve, in other words, that expertise matters.

In Chap. 4 Alenka Poplin and Kavita Vemuri presents an application scenario that is situated in collaborative planning and consensus building, negotiation models, communication in physical vs. digital environment, and Public Participation

Geographic Information Systems (PPGIS). The chapter describes their work to develop a digital spatial game that can enable negotiations about planned urban projects in a large low-income area of Mumbai, India. Their prototype, called **YouPlaceIt!**, was implemented as a web browser-based game with some basic GIS functionality and a satellite imagery base map for orientation. Their goal is to study the implementation of online place-based negotiations and consensus building and the chapter reports some early findings from initial testing with experienced planning professionals. A key observation in this example scenario is the importance of local/regional languages, especially in cultures like India where many distinct regional languages exist. Planning scenarios in urban areas can present especially complex negotiation processes where many native languages could be involved in one negotiation topic/process.

Chapter 5 by Byron Antoniou and Christoph Schlieder describes spatial allocation games as a subclass of location-based games suitable for addressing public participation issues. They study the spatial behavior of contributors to **OpenStreetMap** and links it to gamification mechanisms which provide a solution to issues that arise with patterns of participation. More specifically, three issues are identified: (1) high productive contributors show little commitment to return and update geographic features they created, (2) the gap between the accumulated percentage of created features and the accumulated percentage of updated features is widening, (3) there is a significant contrast between areas of high and low mapping activity. Based on an analysis of the geogames **Geographing**, **Foursquare**, **Ingress**, and **Neocartographer**, six common design patterns for the allocation and deallocation of places are identified. They show how the participation issues map onto the game design patterns, and results from an agent-based spatial simulation provides insights into the interaction of the spatial design pattern.

Chapter 6 presents a second chapter from the same research group. Christoph Schlieder, Dominik Kremer, and Thomas Heinz identify an important part of the geogame design process, namely game relocation, and provide the methodological and technical means for addressing this part of the process in the classroom. While teachers have used geogames in a variety of learning contexts in secondary education, they generally avoid letting the students themselves design the game because of the alleged complexity of the task. In game relocation, the designer adapts a successful geogame to a new geographic environment. The approach taken by Schlieder and colleagues features three components. First, they show how to decompose the game relocation process into a sequence of spatial analysis tasks accessible to students. Second, they present a method, 'place storming', which permits students to search the geographic environment for potential places of game actions. Last, they describe a software tool developed to support students solving the spatial analysis tasks involved in game relocation.

Chapter 7 by Simon Scheider and Peter Kiefer also focuses on game relocation as a core problem in the field of Geogames. While game designers have intuitions helping them to distinguish better from poorer relocations, no concise general quality criteria have been formulated so far. The chapter provides quantitative criteria for relocation of a generic game model and illustrates them with examples. Although the approach uses formal notation, the basic idea can be stated informally. There are essentially two ways in which a relocation can fail and quantitative criteria should capture both of them. First, the environmental embedding can prevent actions, which the game mechanics would permit to take place. Physical barriers preventing access to a place is a typical example. The second type of failure occurs when the interpretation of game actions in terms of physical actions in the environment leads to inconsistent game states. If ownership of a place is obtained just by moving to that place, cases of multiple ownership of a place may occur. In other words, relocation ties narratives to their physical implementation and these ties may be more or less supportive to the rules of the game.

A more critical analysis of the relocation issue is provided by Jim Mathews and Christopher Holden in Chap. 8, where they give an extensive review of existing games that elaborate on the affordances of combining geogaming with place-based education. Their main critique of many geogames to date is that they are often simply "dropped onto" places, without much concern for the local context. Instead they argue for the adoption of place-based education practices in order to design games with an emphasis on learning experiences that are situated within a local community. They also argue for small, locally based development of such games as opposed to large scale game-based curriculum design.

Chapter 9 by Nathaniel Henry is motivated by the relatively large cost to gather very detailed and naturalistic 3-dimensional data from real environments for use in 3D game engines. Henry introduces a workflow that combines low-cost data collection from unmanned aerial vehicles (UAVs), 3D reconstruction methods, and techniques for importing geographic data into a game engine. This approach offers a citizen-centric, low-cost and time-efficient method for Indie geogame designers to capture real terrain for use in a 3D virtual environment where users can navigate from a first-person perspective and view ground objects as they might appear in real life.

While Chap. 10 by Michael Demers does not deal with a geogame per-se, it discusses some important elements of games and learning in a highly geospatial setting. First, following the widely used Quality Matters rubric, it argues for the importance of identifying and aligning course objectives with assessment, instructional material, activities, and technology among other things. Second, under the Quest-Based Learning approach, it describes different forms of gamification and some of the typical game mechanics that are used to gamify a learning activity. At the center is a description of his own development and implementation of a GIS course that follows a quest-based format.

Geographic data and the way they are used in game play characterize geogames. In Chap. 11, the final chapter of the book, Cheng Zhang shows that such data is not necessarily limited (as the prefix geo- suggests) to terrestrial environments. A spatial treasure hunt may take place on the moon if terrain models and other scientific data are available. The basic idea explored by Zhang is quite simple: find a suitable mapping between places on the moon and terrestrial places, which permit to explore lunar caches by walking and searching caches on Earth.

We recognize that the realm of geogames and geoplay is located in the intersection of two rapidly evolving fields: gaming and GI technology. Any work in this area runs the risk of quickly becoming outdated, unless the focus is on underlying principles and theories rather than the technology itself. We think that the contributions in this book emphasizes the former and therefore will provide a lasting reference for future work.

References

Björk, S., Lundgren, S., & Holopainen, J. (2003): Game Design Patterns. In: J. Raessens, M. Copier, J.Goldstein, and F. Mäyrä (eds.): DiGRA '03 - Proceedings of the 2003 DiGRA International Conference: Level Up (pp. 180–193). Utrecht, NL. Digital Game Research Association. Retrieved from http://www.digra.org/wpcontent/uploads/digital-library/05163.15303.pdf

Björk S, Holopainen J (2004) Patterns in game design (game development series). Charles River Media, Newton Centre, MA. http://www.amazon.ca/exec/obidos/redirect?tag=citeulike09-20&path=ASIN/1584503548

Davidsson O, Peitz J, Björk S (2004) Game design patterns for mobile games. Project Report to Nokia Research Center, Finland

Gamma E, Helm R, Johnson R, Vlissides J (1995) Design patterns: elements of reusable object-oriented software. Addison-Wesley, Reading, MA

Holopainen J (2011) Foundations of gameplay. Doctoral dissertation Series No 2011:02. Blekinge Institute of Technology, Karlskrona, Sweden. http://www.diva-portal.org/smash/record.jsf?pid=diva2:835337

Janelle DG, Goodchild MF (2011) Concepts, principles, tools, and challenges in spatially integrated social science. In: The SAGE handbook of GIS and Society. Sage Publications, Thousand Oaks, CA, pp 27–45. https://books-google-com.proxy.lib.ohio-state.edu/books?id=7kuS_P70 YhkC&lpg=PA27&ots=HlarnIqihD&lr&pg=PA27#v=onepa ge&q&f=false

Järvinen A (2008) Games without frontiers: theories and methods for game studies and design. Tampere University Press, Tampere, Finland. http://tampub.uta.fi/handle/10024/67820

Kuhn W (2012) Core concepts of spatial information for transdisciplinary research. Int J Geogr Inf Sci 26(12):2267–2276. https://doi.org/10.1080/13658816.2012.722637

Kuhn W, Ballatore A (2015) Designing a language for spatial computing. In: Bacao F, Santos M, Painho M (eds) AGILE 2015. Springer, Cham, pp 309–326. http://escholarship.org.proxy.lib.ohio-state.edu/uc/item/04q9q6wm

Nijholt A (ed) (2017) Towards playful and playable cities. Springer, Singapore. http://link.springer.com.proxy.lib.ohio-state.edu/chapter/10.1007/978-981-10-1962-3_1

Nyerges TL, Couclelis H, McMaster RB (eds) (2011) The SAGE handbook of GIS and Society. Sage Publications, Thousand Oaks, CA. https://books-google-com.proxy.lib.ohio-state.edu/books/about/The_SAGE_Handbook_of_GIS_and_Society.html?id=7kuS_P70YhkC

Nyerges TL, Karwan M, Laurini R, Egenhofer MJ (eds) (1995) Cognitive aspects of human-computer interaction for geographic information systems, vol 83. Kluwer Academic Publishers, Netherlands. https://books-google-com.proxy.lib.ohio-state.edu/books/about/Cognitive_Aspects_of_Human_Computer_Inte.html?id=cePdBgA AQBAJ

Sintoris C (2015) Extracting game design patterns from game design workshops. Int J Intell Eng Inform 3(2–3):166–185. https://doi.org/10.1504/IJIEI.2015.069878

Chapter 2
Defining a Geogame Genre Using Core Concepts of Games, Play, and Geographic Information and Thinking

Ola Ahlqvist, Swaroop Joshi, Rohan Benkar, Kiril Vatev, Rajiv Ramnath, Andrew Heckler, and Neelam Soundarajan

2.1   Introduction

In 2008, the National Science Foundation (NSF) released the report "Fostering Learning in the Networked World: The Cyberlearning Opportunity and Challenge". NSF argued in this report that the heavy investment and focus on Cyberinfrastructures must be complemented by a parallel investment in Cyberlearning, "…learning that is mediated by networked computing and communications technologies." (Borgman et al. 2008). The rationale was that information and communication technologies had reached a critical tipping point where high-end computing, cyberinfrastructures and mobile technologies were readily available for billions of users, but it was still unclear what affordances they could bring to learning in structured classroom settings and more informal learning environments.

As a consequence of the report recommendations, NSF created the Cyberlearning program within the Division of Information and Intelligent Systems (IIS) directorate which began funding a broad collection of projects, ranging from the exploration of new ideas to project implementation and scaling-up of thoroughly tested technologies. A separate Center for Innovative Research in CyberLearning (CIRCL) was also established to support and amplify the funded projects, and the Center website http://circlcenter.org provides plenty of information on the ongoing research.

A common theme found in these projects is the desire to help leverage new cyber-technologies for learning. Important questions are raised: what are entirely

O. Ahlqvist (✉)
Department of Geography, The Ohio State University, 1049B Derby Hall, 154 N Oval Mall, Columbus, OH 43210, USA
e-mail: ahlqvist.1@osu.edu

S. Joshi • R. Benkar • K. Vatev • R. Ramnath • A. Heckler • N. Soundarajan
Ohio State University, Columbus, OH 43210, USA

© Springer International Publishing Switzerland 2018
O. Ahlqvist, C. Schlieder (eds.), *Geogames and Geoplay*, Advances in Geographic Information Science, https://doi.org/10.1007/978-3-319-22774-0_2

new opportunities that these cyberinfrastructures provide that could support learning? How does learning happen with these technologies? Are there generalizable theories that could support designers of future cyberlearning environments?

The work we report on in this chapter was funded by the NSF Cyberlearning program as an Exploration (EXP) project entitled "GeoGames—A Virtual Simulation Workbench for Teaching and Learning through a Real-World Spatial Perspective". The project ran from 2011 to 2016 and is now in its concluding phase.

2.2 The Genesis of Our Geographic Information Systems-Multiplayer Online Games (GIS-MOG) Idea

Our venture into the realm of combining online maps with board games came in the spring of 2007 during an independent study between the lead author and an undergraduate geography student who wanted to design a board game map for the popular game Ticket to RideTM by Days of Wonder. The game objective is to connect cities with railroad lines by collecting cards of various types that can be combined to claim segments along prescribed railway routes. While the original board game was played on a game map of the United States, the popularity of the game ensured that versions for other parts of the world was developed, both as official versions released by the publisher Days of Wonder, and also as unofficial maps created by an active user community.

The independent study assignment was to design a new game map of Canada. As this work progressed, there were repeated conversations around what type of geographic insights board games like Ticket to RideTM and other popular map-based games could provide. Clearly, many board games do make use of some type of map that can provide various types of geographic learning depending on what is emphasized either through the map itself or other game elements and mechanics that take some grounding in geographic reality.

While games for geographic learning have existed for a long time, Walford (1981) noted in his review of developments from the 1960s to the 1990s that there was a total lack of systematic evaluation of the learning experiences and outcomes of such games. In the years since, there have been some efforts to that end, but much of what is learned from those studies are typically hard to generalize beyond a particular game and its specific educational context. It is worth noting that this issue is not unique among geographic education studies. The CyberLearning program came as a response to similar concerns about education research across many domains, and the program was specifically interested in projects that would be able to create generalizable knowledge about how and why learning happens with cybertechnology (Borgman et al. 2008). Consequently, in addition to better understand the opportunities and obstacles presented by our GIS-MOG framework for role-play games/simulations, we started an iterative design process to develop an example prototype application and explore the affordances of this new technology platform that uses GIS maps as game boards for geographic learning. Central research questions were: what does this new learning technology provide in terms of authentic experiences, student engagement, and higher-order thinking? And how do the specific technology affordances (access to rich geographic information, particular game mechanics, collaboration opportunities, etc) help or hinder learning? Answers to these two questions will be reported elsewhere. In this chapter we will focus on a third question that we were equally interested in: what is our prototype geogame application representative of, as a broader category, or genre, of learning technologies? Describing the game by specifying key components and functionality can help and guide others to develop similar technology, compare across different implementations, and hopefully allow for the generalization of findings across independent studies of specimen applications.

Before getting to this description, we will first review how our particular technology for a geogame prototype iteratively evolved, each version incorporating key features of the new learning technology.

2.3 The GIS-MOG Technology Framework

In our technology development work we followed a design based research pattern (Barab and Squire 2004) by progressing through a series of five major iterations to incrementally build from our original idea to the current proof-of-concept prototype.

Our first iteration (Fig. 2.1a) was carried out in 2008 to initially explore real-time interactions in an online mapping environment. At that time, no existing online mapping environment provided such functionality so a custom architecture was built using Openlayers (www.openlayers.org), PHP and MySQL. This demonstrated the viability of a light-weight client solution with cross platform compatibility and near instantaneous (<1s) propagation of user actions between clients. In the next iteration (Fig. 2.1b) we used the first release of the Google Earth browser plugin and associated API to support inclusion of distributed and custom geospatial data in a command and conquer game styled after the board game RISK, while maintaining real-time multi-user interaction through the map interface.

In the third iteration (Fig. 2.1c) we introduced the support for real-time web-services such as the WMS, WFS, and WPS implementation specifications (see http://www.opengeospatial.org/standards), to demonstrate how this could change the conditions of any new game by providing ever changing condition variables e.g. weather, stock markets, house prices. The next iteration (Fig. 2.1d) sought to create a fully playable prototype for user testing by expanding capabilities to interaction between players through a game lobby, chat room, and discussion board. This platform also implemented all capabilities from previous iterations through an integration of two JAVA based client-application frameworks; NASA's World Wind (http://worldwind.arc.nasa.gov/) virtual globe environment and Sun's Massive Multi-Player Online gaming platform **Darkstar**. Throughout the development the games were play-tested by researchers (iterations a–c), small focus groups (iteration c and d) and entire class sections (iteration d) at the Ohio State University. For more information about these stages in our development see Ahlqvist et al. (2012).

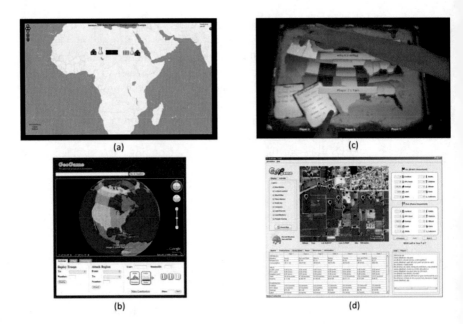

Fig. 2.1 The first four major iterations of the GIS-MOG platform. (**a**) Using Openlayers for real-time multi-user interaction. (**b**) Using Google Earth JavaScript API for geodata integration. (**c**) Using web-services for live geodata feeds. (**d**) Using World Wind and Darkstar for fully playable integration of geodata and real-time multi-player support

After the completion of the four experimental prototypes a–d, we identified that the combination of web-based GIS and multi-player game mechanics seemed to offer a novel and powerful learning technology with a unique combination of desired affordances. With these insights we embarked on a longer term design experiment, with the goals of both building a fully GIS-based map game environment and doing repeated experiments to study how and why learning can happen with this novel technology.

Our current prototype game lets students in an introductory Geography course explore the concept of the Green Revolution from the perspective of a farmer in rural Punjab, India. After a signup process, students get to choose a family and start farming on digital plots of land located in one of the many farming villages in that region. The main game interface (Fig. 2.2) presents users with an aerial photo map of a village in Punjab, where plots of land that players can acquire and farm are outlined on top of the aerial image.

The game is turn-based and each turn represents a single growing season. During a turn, players can interact with the map by clicking on parcels to identify the size and cost of buying that piece of land. Once owned, a parcel is open for various farming options (Fig. 2.2, #1–3): players can plant parcels with land race or high-yield seeds (#1), apply fertilization at low or high levels (#2), and irrigate the fields after first investing in an irrigation system (#3) that will have varying cost depending on

Fig. 2.2 The GIS-MOG game play interface

how far away from the river a parcel is. The players have to decide on how to manage each owned parcel before a round is ended (Fig. 2.2, #6). After a turn ends, the yield from each parcel is calculated by a set formula that accounts for the chosen farming choices and a random weather index between 1 (good) and 5 (poor). After the yield has been calculated players can buy and sell any surplus yield (at a randomly set market price) using the market tab (Fig. 2.2, #5), or keep it for future seasons. The next round then ensues where players again have to decide on planting, fertilization, irrigation and possibly buying more land. The game ends after a set number of rounds and winners may be determined based on the value of accumulated assets.

2.4 Defining the GIS-MOG Genre

As mentioned above, one guiding question in our research has been to identify how to classify the GIS-MOG game as a more general category or genre of games and learning technologies. It is common to find groupings of games into so-called 'genres', for example puzzles, role-play, simulation, sports and more, but until recently these sub-divisions were rather arbitrary and typically not helpful for efforts to generalize research findings from one game to others in the same genre. Building on work by genre theorist Rick Altman, Järvinen (2008) notes in a lengthy analysis of game genres that genres are not stable constructs but evolve over time; A genre is comprised of not just systemic (components of the game system itself) but also thematic, contextual, and rhetorical aspects. In our case, we were developing a very particular game technology motivated by new opportunities in two separate technological realms: Geographic Information Systems (GIS) and Massively

Multiplayer Online Games (MMOG). In the first realm, we identified increasingly cyber-enabled GIS as an emerging paradigm for the provision of access to almost unlimited and multi-faceted information about the world, and powerful and scalable geoprocessing capabilities for analysis, modeling, and simulation of real world processes (Wang 2010; Wright and Wang 2011). In the second realm, we found MMOG as a rapidly emerging space where highly immersive, graphically rich videogames were delivered and played online by large numbers of individuals (c.f. Chan and Vorderer 2006). In order to properly situate and characterize our new learning technology that integrate both realms, we will consider the key geographic concepts and game mechanics, elements, and design patterns identified in the introductory chapter (Ahlqvist and Schlieder, Chap. 1).

The technology configuration we have developed in this project is an integration of state-of-the-art GIS and online multi-player game technologies. It could be thought of as a board game that is played on top of the online, interactive maps that you find on the web, e.g. Google and Bing maps. The fundamental innovation of the GIS-MOG platform is the ability to design games on a modifiable and interactive geographic game board. The platform integrates a full range of GIS-supported map and processing services with online multi-player gaming affordances, combined into an online map game/simulation environment.

As we saw in the introductory chapter, there has been a growing interest from academics to define the key characteristics of game genres, including Location-Based Games. Some of the existing genre labels, like pervasive games, ubiquitous games, augmented-, alternate-, and mixed-reality games, mobile games, geogames, and adaptronic games, are often tech-centric, meaning that the terminology would have to constantly change with changing technology (Holopainen 2011, p. 30). Our review in Chap. 1 identified some key geographic concepts and game patterns/elements from the literature that are largely technology agnostic, yet represent a best effort at describing fundamental concepts in each domain. We also demonstrated (Ibid.) a strong alignment between many of these key concepts. While these collections of core concepts are nascent, they hold promise for helping to pin down and create structure in what is currently a rapidly expanding number of heterogeneous geogame and geoplay applications and activities. We invite the reader to review Chap. 1 for an overview of the identified geographic concepts, game patterns and game elements.

Concurrently, and particularly relevant to this chapter, there have been several efforts to categorize learning technologies (Culatta 2011). One direction that learning research has taken is to describe learning technology in terms of Learning Objects, defined by Churchill (2007) as a representation designed to afford uses in different educational contexts, consisting of six object sub-types: **presentation, practice, simulation, conceptual models, information,** and **contextual representation** objects. Recognizing the increasing integration of learning technology with traditional learning approaches, Graham (2005) identified four key dimensions of learning technologies—**Space, Time, Fidelity** and **Humanness**—to describe how certain affordances are enabled through the learning technology. The Space dimension describes to what degree real life is mixed with virtual reality.

The Time dimension describes the immediacy of interactions from real-time to asynchronous interactions with longer lag time. The Fidelity dimension describes the sensory richness of a technology, from involving all senses to only one, such as text only. Finally, the Humanness dimension describes the degree to which the learning technology is part of the learning experience, from more or less absent to being entirely mediated through the technology.

In the following we will present a characterization of our GIS-MOG technology using the extended set of core geographic concepts from the introductory chapter, Table 1.5. The presentation is organized using a framework provided by Ahlqvist (2017) who identified *Representation, Spatial and Temporal Expansion, Location,* and *Pervasiveness* as key dimensions of location-based games. This division has many similarities to the learning technology categories above, but it was developed with a focus on location-based games. Section 2.4.1 considers the way that the game represents (most often digitally) real space and time in an abstract, digital information environment. Section 2.4.2 discusses how location in space-time determines the game systems dynamics. Sections 2.4.3 and 2.4.4 considers the way space and time is scaled, expanded, or compressed. Finally, Sect. 2.4.5 considers the degree to which the game allows participants to move between real and represented environments.

By describing these four dimensions, using the extended set of core geographic concepts from Chap. 1 (identified in bold face below) and examples from our own geogame prototype, we posit that identifying and classifying the key components, functionality, and affordances of the GIS-MOG framework will help others to develop similar technology, make comparisons, and draw inferences between different geogames.

2.4.1 Representation

The representation of geographic information is a long-standing issue that has generated significant academic debates over the past half century or so (Fisher and Unwin 2005). Our GIS-MOG framework allows any representation currently supported by ArcGIS (e.g. feature layer, raster, network) to be part of the game interface. This means that we can use most GIS information (maps, networks, remote sensing imagery, etc.) of the real world for game play. For example, we may choose to incorporate data on climate, soils, water resources, demography, economy, weather, traffic, and other geographic themes, as well as satellite and aerial imagery, Digital Elevation Models, and even dynamic data from real-time weather stations, traffic monitors or social media feeds. As such, GIS-MOGs are agnostic to any specific instantiation of the core geographic concepts of **Object**, **Field**, and **Network**. The GIS-MOG framework is capable of implementing several different representational formats such as square grids, hexagonal grids, Triangulated Irregular Networks, as well as Point, Line, and Area vector objects, with support for various import and export formats.

The simulation and gaming aspect of a GIS-MOG scenario requires that represented game space and time are augmented with game items that use the same core concepts above but only exist in the game world and do not correspond to real world features. Examples of these are objects such as player avatars, non-player characters, tokens, field-type attributes that give certain parts of the environment some type of value in the game, or connections (networks) like virtual portals that allow for jumps across space or time (see Sect. 2.4.4 about space/time warping). This combination of real world and game world is essentially a Hybrid Space (Davidsson et al. 2004) game pattern.

An important concept related to objects that exert some kind of behavior is **Agency**. Players and game facilitators are obvious examples of game agents that makes decisions and triggers game actions, but agency can also be coded into other game objects as well as the game environment representations. As a geographic concept, agency has emerged as a key concept in order to understand world systems as a combination of spatial structure and agency at different scales (c.f. Flint and Shelley 1996; McLaughlin and Dietz 2008). In a GIS-MOG context, agency will be expressed by the players and possible to embed with other game elements through rules and associated simulations (see Sect. 2.4.4 below.)

In our current GIS-MOG we have used actual remote sensing imagery serviced form the Esri World Imagery map service (Esri 2017b). This real-world foundation is used as a source for augmenting the game world with a "farm land" feature layer consisting of digitized land plots that roughly correspond to real agricultural plot boundaries surrounding the village on the ground, and a "River" network layer that is digitized from the imagery to roughly match up with an existing river that runs by the village. Each of these layers has added attributes that determine some of the game mechanics.

2.4.2 Location

The driving idea behind our GIS-MOG framework is that location in the game is important in the same way that location is important in geography. Space-time location and analysis is at the heart of GIS data management and operations, which means that our GIS-MOG framework is inherently maintaining location information for all game environments/elements through one or more of the Field, Object, and Network representations. Location in the generic sense is always a relation between a figure and some chosen ground (Kuhn 2012). In most games that ground is typically an internal reference system, game board, etc., with no direct correspondence in the real world. In a GIS context the ground is typically some chosen geographic reference system that allows for a direct correspondence between represented features in the system and a true location somewhere in the world. As a consequence, any aspect of the space-time location of GIS-MOG elements, such as players, avatars, tokens, other game objects, and the overall game environment, can be informed by additional information form the real world about that location, and which may in turn affect the game dynamics. This dynamism is central to our GIS-MOG framework.

Standard spatial analysis functions in GIS can perform various types of **Distance** and **Neighborhood** operations, as well as a wide variety of **Overlay**, **Spatial heterogeneity**, and **Spatial dependence** operations. These analytical procedures can be used to determine a wide range of possible interactions between game objects and the game environment. These core concepts correspond directly with the locality and proximity game patterns identified in the introductory chapter. Our current GIS-MOG implements a few of these analytical concepts to determine the cost of parcels (area and distance to river), cost to install irrigation (distance to river), water availability (more water upstream in the river network), and neighborhood/connectivity (adjacent land can be irrigated by the same installation). Another possibility that has been considered for our current game is varying the land parcel cost and yield depending on soil quality (Overlay). Most of the analytics necessary to determine those rules and dynamics are inherently supported by the GIS back-end of our technology.

2.4.3 Spatial Expansion

Spatial **scale** is a central notion in geography with many meanings. Some of the most important notions are scale as a way to define the spatial extent of a study, the operational scale of spatial phenomena, the degree of detail in spatial information, and the cartographic representational fraction that defines the correspondence between measurements on a map with real measurements on the ground. Despite its fundamental role and a relatively well-defined concept, scale remains a surprisingly active area of geographic research and inquiry (Sheppard and McMaster 2004). In the context of serving as a defining characteristic for geogames as a learning technology, it is particularly relevant to consider perceptual and cognitive scales to specify the spatial extent of a geogame.

Montello (1993) identified figural, vista, environmental, and geographical space as four cognitively distinct scale-ranges that are qualitatively different in how humans treat and understand them. In our GIS-MOG framework the game environment expands play beyond a room, or a soccer field, to larger geographic spaces like villages, cities, countries and continents. Cognitively, this would correspond to the environmental and geographical spaces. These are too large to be visually apprehended without significant movement and integration of information over time. Consequently, these spaces are particularly amenable to be comprehended through maps or aerial imagery with symbolic or otherwise abstracted representations of reality (see Sect. 2.4.1 above). This is accomplished by scaling the real space down, in a traceable way that maintains a real world connection, to a manageable size (e.g. computer screen or table-top board) for the purpose of game play.

With this restriction it makes our GIS-MOG framework distinct from games with motion-control (e.g. Wii and Kinect) that are played in figural space, and games that are played in vista space like **Pac Manhattan** and **OriGami**, where vision, haptics, head and eye movements are primary sensorimotor systems for interaction with and understanding the game.

In later work, Montello and Raubal (2013) proposed that there are at least six, partially overlapping categories of spatial-cognitive tasks that people perform regularly and in varying cognitive scale-ranges/spaces. Because our GIS-MOG framework is entirely mediated through a screen interface with no need for the user to navigate or experience the real world in order to play, up to four of these tasks are involved; "Using spatially iconic symbolic representations", "Using spatial language", "Imagining places and reasoning with mental models", and "Location allocation". The remaining two categories, "Wayfinding as part of navigation" and "Acquiring and using spatial knowledge from direct experience", requires direct physical interaction with the geographic space and is more associated with previously mentioned location-based games that take place in figural and vista spaces.

2.4.4   Temporal Expansion

Time can be scaled up or down in order for the game play to span and represent longer or shorter time intervals than the time it takes to play the game. Some games may even take on a less determined temporal scope and continue even when you sleep or go to work. The part of our GIS-MOG framework that probably required most attention, as it was least supported by existing GI-systems, was the temporal dimension of game play. As described in Chap. 1, the game literature has a rich vocabulary around the actual game play and associated events (actions, events, closures, game time, game mechanics, etc.)

Based on the definition by Clark et al. (2009), geographic **simulations** would be defined as computational models of real or hypothesized geographic situations or phenomena that allow users to explore the implications of manipulating or modifying parameters within the model. In the Geogame context it is also helpful to distinguish between the separate but interconnected concepts of model and simulation. We think of a geographic model as the digital representation of a real-world phenomenon while simulation is the software framework or architecture within which a model is animated. In this we align with the general definition of a simulation as "…dynamic computer models that allow users to explore the implications of manipulating or modifying parameters within them." (National Research Council 2011 p. 2). Without going into more detail about what separates a game from a simulation, we follow Salen and Zimmerman (2004) "a system is a set of parts that interrelate to form a complex whole" and Järvinen (2008) to view all games as systems. Viewed this way, a Geogame is a model of a geographic system that, when played, enacts a simulation of that geographic system.

A key feature of our GIS-MOG framework, and one that we argue is key to define this particular genre of games, is the integration of computational social or environmental models that can simulate how the geographic (game) system will respond to user actions. As an example, our current farming game implements a simple surface water network model to simulate what will happen to water flow downstream when a player (farmer) in the game starts using river water for irrigation. The system can calculate, based on where in the river network an irrigation system is installed, amount of water flow in the river, and amount of water used for irrigation, how much water is left downstream from the irrigation point. By implementing this model using the framework of stand-alone geoprocessing services (Esri 2017a), we seek to generalize the spatial system components so that each sub-system can be modelled separately and simulated as a sub-component of the entire game system. In doing so, we allow for other designers to author and implement other system behaviors by adding or modifying such models. An example could be designing, authoring and publishing a simple economic model that can be consumed by a GIS-MOG to simulate market prices as an effect of the aggregated farm yield in the game and the supply/demand from other parameters in the game. As such, the GeoGame framework is maybe unique among most other gaming systems in that it allows for the outsourcing of game processes to other, third-party and stand-alone services.

In this context it is worth noting that most geographic models and simulations are typically aiming at a truthful representation of reality. For example, most people would generally expect that a map shows the true location of roads, cities, rivers etc. In games and game simulation we find a more mixed set of priorities and the focus is often more on the entertainment, imaginary aspects, and about the activity itself (Clark et al. 2009). Certainly, many games embed a certain degree of realism as part of the intrinsic features, but not to the degree that geographic simulations aspire to. Similar to space, it is also common that game time is scaled and warped to allow for game play beyond the real time span e.g. to play through 1 year/decade/century in a short game round, or to do particular sequencing, jumps, and loops in space or time to support particular game rules and dynamics. Our green revolution GIS-MOG compresses time so that one growing season becomes one game round, and game rounds are determined either by users triggering the next round manually or by a desired timer, often set to less than 5 min if the game is played in real time.

An **Event** as a core geographic information concept is defined by Kuhn (2012) as a change to location, neighborhood, field, object and network. It is typically carved out as a discrete and temporally bounded chunk from continuous processes. In a gaming context, we can think of the game session as the process from which we can identify specific game events. Tangible events can happen as a result of user actions (e.g. moving or manipulating a spatial representation) or system procedures (e.g. turn change). The temporal components identified by Holopainen (2011) suggest additional gaming/simulation-specific event semantics related to play sessions, set-up and set-down. Yet, for the purposes of this chapter, it is unclear in what way a game is best characterized in terms of how it handles events.

A simulation is ultimately governed by **Rules** that specify how for example water dynamics are calculated. Earlier we described how irrigation costs and the cost of buying a parcel are determined by distance and area measures. With a game defined as a system of interrelated parts, the rules are at the heart of defining and regulating how game elements can and will interact (Järvinen 2008 p. 30). As such, the setting up of rules in a geographic system design has many similarities with a long tradition of research on expert systems and artificial intelligence (c.f. Robinson and Frank 1987). There are obviously many ways to determine rules for geogames. We could base rules on empirically tested and verified physical and social dynamics, for example by implementing a well-researched surface water model. It is also possible to develop rules by eliciting knowledge form experts, farmers, or other stakeholders (c.f. Barreteau et al. 2007) who may have knowledge about the workings of a particular dynamic that a game rule seeks to model. Ultimately, these rules need to be defined, coded into the game and communicated with players.

In our current system, many of the rules are *embodied* (Järvinen 2008) by the game elements themselves, meaning that the way the game and the user **Interface** is designed will determine significant workings of a game. In our game, despite being played using a browser-based online map that in theory can be navigated to any place in the world, the game environment is restricted to one part of a village in rural Punjab because of the parcel layer limits. It is relatively easy to expand the "game board" by adding more parcels, or set up the game in a different place by digitizing a new set of parcel boundaries on top of the map imagery, but that layer very much determines the boundaries of the game in terms of the number of parcels available, their spatial configuration, sizes, etc. The interface itself also sets limits on how a user can access and control the game, notably through points, clicks and text entry in the browser window.

Because the current GIS-MOG is a proof-of-concept prototype, many rules that are specific to the particular Green Revolution game have been hard coded, whereas other rules are expressed as modifiable parameters (variables) in the game code. For this, the GIS-MOG framework uses Web Rule, an XML-based ASP.NET and MVC business rules engine that is accessed through a game administrator interface as part of the GIS-MOG system (Fig. 2.3).

Ultimately, to allow for as much flexibility as possible, most rules would benefit from being modifiable through this interface, but this would amount to also accounting for how to resolve some of the rules that are 'embedded' in the actual game interface. As an example, if we were to add a new type of goods that could be traded through the market interface (see #5 in Fig. 2.2), there would have to be a way to automatically, or at least in some easy way, edit the visual interface to include a new icon and organize the display in a functioning way. If we were to change the type of crops to grow, the way fertilization and irrigation is done, or add other possible farming actions, this would also have direct impact on the interface design (see #1–3 in Fig. 2.2). Clearly, this type of flexibility is far beyond the scope of our project, but this example helps to illustrate how embedded certain game rules are with the interface design.

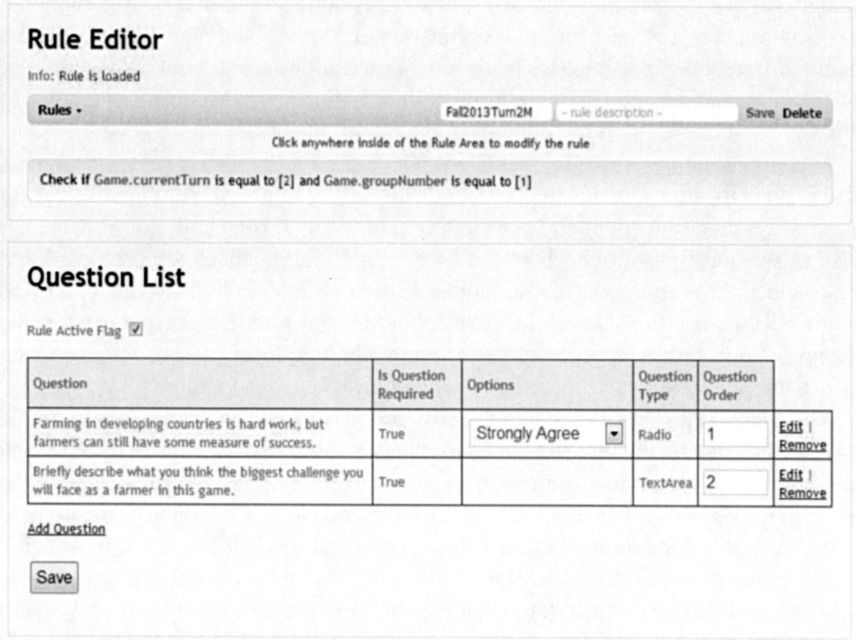

Rule Editor

Info: Rule is loaded

| Rules ▾ | Fall2013Turn2M | - rule description - | Save Delete |

Click anywhere inside of the Rule Area to modify the rule

Check if Game.currentTurn **is equal to** [2] **and** Game.groupNumber **is equal to** [1]

Question List

Rule Active Flag ☑

Question	Is Question Required	Options	Question Type	Question Order	
Farming in developing countries is hard work, but farmers can still have some measure of success.	True	Strongly Agree ▾	Radio	1	Edit	Remove
Briefly describe what you think the biggest challenge you will face as a farmer in this game.	True		TextArea	2	Edit	Remove

Add Question

Save

Fig. 2.3 The Question Configuration Page in the game administrator interface, showing the criteria in the Rule Editor that triggers particular questions. Those questions are entered through the Question List editor

2.4.5   Pervasiveness

Being primarily concerned with location-based games, Ahlqvist 2017) included Pervasiveness as a key dimension to describe the degree to which the game allows participants to become immersed in the simulation, to move between real and represented environments, and to infer meaning and value from the game. **Interfaces** such as mobile technology, wearable computers, head-mounted displays, sensor networks and other pervasive computing technologies can allow a geogame to mix in with the real world such that the boundary between what is part of the game and not is blurred. However, since the real world is infinitely complex, save for man-made artifacts where each component is entirely known, any representation inevitably causes some amount of abstraction and generalization of the real world. As a way to specify the level of abstraction, geographic representations can most typically be further specified in terms of their **Granularity** (e.g. resolution, minimum mapping unit) and various aspects of **Accuracy** (e.g. spatial, temporal) in order to determine the degree of detail that is represented. These specifications are either an inherent feature of the previous representation concepts (e.g. the resolution of a raster data set) or a separate but complementary feature (e.g. Root mean square error estimates of positional accuracy as part of metadata). Semantic accuracy

(Salge 1995) specifications and using ontologies and folksonomies for geographic information can provide a formal specification of a "perceived reality" that can reconcile different perspectives. Rich semantic descriptions are still not a standard part of spatial metadata, yet it provides critical information about **Meaning** as it helps to answer questions about how to interpret the representations.

It is important to remember that GIS-MOG games make it easy for the designers and players to situate games in places and cultures far removed from themselves. In some ways this is a benefit and even a reason for using this technology. However, it also raises questions around how our own values and practices are promoted and reproduced through the design and game play. As Mathews and Holden (Chap. 8) point out, we need to involve local stakeholders and multiple perspectives in the design of our games in order to better reconcile how local places, societies, and issues are represented and remediated when players engage in game play.

Ultimately, the degree to which the abstractions and interface manages to mimic how we understand the real world is important for how players will engage with and understand the game environment. Yet, many abstract 'game worlds', such as the grids in the classic games of Tic-tac-toe, Chess, and Go, convey enough meaning to generate highly captivating games. This suggests that it is feasible to use more schematic representations in GIS-MOGs and still produce an engaging experience. Examples of this could be a game that uses subway maps or cartograms that maintain some tractable transformation of the real world into the represented game environment.

2.5   Summary and Discussion

Through this overview we have sought to provide a rich and multi-faceted description and definition of a new geogame learning technology genre called Geographic Information Systems-Multiplayer Online Games (GIS-MOG). We did so by discussing how GIS-MOG incorporates 18 (Chap. 1, Table 1.4) core concepts related to games, play, and geographic information and thinking.

Being built on a GIS foundation, the GIS-MOG framework allows for a variety of game world representations, including most existing spatial data formats supporting object, field and network structures (Kuhn 2012). As a consequence, the GIS-MOG framework is characterized by a flexible and rich array of location-based information and analytic functionality able to support most locality and proximity game patterns (Ahlqvist and Schlieder, Chap. 1). Our GIS-MOG genre is however prescriptive about the spatial game scale as it primarily engages with environmental and geographic cognitive spaces (Montello 1993), and as a result it is primarily concerned with game activities related to the use of spatial language, spatially iconic symbolic representations, imagining and reasoning about places with mental models, and conducting location-allocation activities (Montello and Raubal 2013). In terms of temporal expansion, the GIS-MOG platform implements all temporal components identified by Holopainen (2011), and it implements a rule set that is not necessarily prescriptive but constrained by the user interface which is primarily a device screen with input mechanisms such as a keyboard and pointing device. Another distinguishing feature of our GIS-MOG genre is the possibility to outsource game functionality to geoprocessing services outside the core game software as part of the simulation and rules, offering a way to build flexible and modular simulations. Game world data in a GIS-MOG either have inherent or supplemental metadata that accounts for the accuracy and granularity of game representations. Probably the most difficult dimension to account for is the pervasiveness as it includes aspects of interface, meaning and value.

Our current farming game is one instance of an infinite number of games that can be implemented on our GIS-MOG platform. We see a big potential with the possibility for game designers to customize a game, for example by moving the location of an existing farm management game to their own neighborhood, modify the factors involved, or changing some rules to create a new and unique simulation experience. In our most current work we have experimented with a first iteration of what we call a "Game Builder" interface that will allow an administrator to move the current game to any location in the world, or travel back in time using historical maps and data to 're-live' an illustrative historical example, with only a few inputs from the administrator. We also see exciting opportunities in studying the game system, including the decisions made by players. Through the game we get a unique window into aspects of geographical decision making that is hard, if not impossible, to gain by just watching real world geographic systems.

Implementing multi-player gaming support, the GIS-MOG allows for distributed collaboration between anything from two to thousands of users on the simulation platform. What this means is that we offer an environment that has many similarities with **SimCity**, but where we use authentic, real-world geography and capability to make modifications. We are confident that there are other systems, existing or under development, that have very or somewhat similar characteristics as our geogame. Our hope is that a detailed description like the one we provided in this chapter can help users and designers to identify key similarities and differences that can help guide their use and design of geogame technologies, and compare experiences and findings. We recognize that this is a first-of-its-kind effort to do such a structured description of a geogame technology and we look forward to see further development of the set of descriptive criteria as well the informed use of descriptions like these. Nevertheless, we feel confident that we have investigated this technology to the point that can be considered representative of a unique yet broad genre of learning technology.

References

Ahlqvist O, Loffing T, Ramanathan J, Kocher A (2012) Geospatial human-environment simulation through integration of massive multiplayer online games and geographic information systems. Trans GIS 16(3):331–350. https://doi.org/10.1111/j.1467-9671.2012.01340.x

Ahlqvist, O. (2017). Location-Based Games. In International Encyclopedia of Geography: People, the Earth, Environment and Technology (pp. 1–4). John Wiley & Sons, Ltd. https://doi.org/10.1002/9781118786352.wbieg0298

Barab S, Squire K (2004) Introduction: design-based research: putting a stake in the ground. J Learn Sci 13(1):1–14

Barreteau O, Le Page C, Perez P (2007) Contribution of simulation and gaming to natural resource management issues: an introduction. Simul Gaming 38(2):185

Borgman CL, Abelson H, Dirks L, Johnson R, Koedinger KR, Linn MC, Lynch CA, Oblinger DG, Pea RD, Salen K, Smith MS, Szalay A (2008) Fostering learning in the networked world: The cyberlearning opportunity and challenge. National Science Foundation, Virginia, p 12. www. nsf.gov/pubs/2008/nsf08204/nsf08204.pdf

Chan E, Vorderer P (2006) Massively multiplayer online games. In: Vorderer P, Bryant J (eds) Playing video games: motives, responses, and consequences. Lawrence Erlbaum Associates Publishers, Mahwah, NJ, pp 77–88

Churchill D (2007) Towards a useful classification of learning objects. Educ Technol Res Dev 55(5):479–497. https://doi.org/10.1007/s11423-006-9000-y

Clark D, Nelson B, Sengupta P, D'Angelo C (2009) Rethinking science learning through digital games and simulations: genres, examples, and evidence. In: Learning science: computer games, simulations, and education workshop sponsored by the National Academy of Sciences, Washington, DC. http://sites.nationalacademies.org.proxy.lib.ohio-state.edu/cs/groups/dbassesite/documents/webpage/dbasse_080068.pdf

Culatta R (2011) Taxonomies of learning technologies. http://innovativelearning.com/instructional_technology/categories.html. Retrieved 29 Jan 2017

Davidsson O, Peitz J, Björk S (2004) Game design patterns for mobile games. Project Report to Nokia Research Center, Finland

Esri (2017a) What is a geoprocessing service?—Documentation. ArcGIS Enterprise [Software documentation]. http://server.arcgis.com/en/server/latest/publish-services/windows/what-is-a-geoprocessing-service-htm. Retrieved 20 Feb 2017

Esri (2017b) World imagery. https://www.arcgis.com/home/item.html?id=10df2279f9684e4a9f6a 7f08febac2a9. Retrieved 23 Feb 2017

Fisher, P. F., & Unwin, D. J. (Eds.). (2005). Re-presenting GIS. John Wiley & Sons

Flint C, Shelley F (1996) Structure, agency, and context: the contributions of geography to world-systems analysis. Sociol Inq 66(4):496–508

Graham C (2005) Blended learning systems: definition, current trends, and future directions. In The handbook of blended learning: global perspectives, local designs. Wiley, New York, p 624

Holopainen J (2011) Foundations of gameplay. Doctoral dissertation Series No 2011:02. Blekinge Institute of Technology, Karlskrona, Sweden. http://www.diva-portal.org/smash/record. jsf?pid=diva2:835337

Järvinen A (2008) Games without frontiers: theories and methods for game studies and design. Tampere University Press, Tampere, Finland. http://tampub.uta.fi/handle/10024/67820

Kuhn, W. (2012). Core concepts of spatial information for transdisciplinary research. International Journal of Geographical Information Science, 26(12), 2267–2276. https://doi.org/10.1080/13 658816.2012.722637

McLaughlin P, Dietz T (2008) Structure, agency and environment: toward an integrated perspective on vulnerability. Glob Environ Chang 18(1):99–111. https://doi.org/10.1016/j. gloenvcha.2007.05.003

Montello DR (1993) Scale and multiple psychologies of space. In: Frank AU, Campari I (eds) Spatial information theory: a theoretical basis for GIS, vol 716. Springer-Verlag, Berlin, pp 312–321

Montello DR, Raubal M (2013) Functions and applications of spatial cognition. In: Waller D, Nadel L (eds) Handbook of spatial cognition. American Psychological Association, Washington, DC, pp 249–264. https://doi.org/10.1037/13936-014

National Research Council (2011) In: Honey M, Hilton M (eds) Learning science through computer games and simulations. The National Academies Press, Washington, DC. http://www. nap.edu/catalog.php?record_id=13078

Robinson VB, Frank AU (1987) Expert systems for geographic information systems. Photogramm Eng Remote Sensing 53(10):1435–1441

Salen K, Zimmerman E (2004) Rules of play: game design fundamentals. MIT Press, Cambridge, MA

Salge F (1995) Semantic accuracy. In: Guptill SC, Morisson JL (eds) Elements of spatial data quality, vol 1. Elsevier Science Ltd., Oxford, pp 139–151

Sheppard E, McMaster RB (eds) (2004) Scale and geographic inquiry. Wiley, New York. https://books-google-com.proxy.lib.ohio-state.edu/books/about/Scale_and_Geographic_Inquiry.html?id=vJiRSeXJ-soC

Walford, R. (1981). Geography games and simulations: learning through experience. Journal of Geography in Higher Education, 5(2), 113–119. https://doi.org/10.1080/03098268108708808

Wang S (2010) A CyberGIS framework for the synthesis of cyberinfrastructure, GIS, and spatial analysis. Ann Assoc Am Geogr 100(3):535–557

Wright DJ, Wang S (2011) The emergence of spatial cyberinfrastructure. Proc Natl Acad Sci U S A 108(14):5488–5491. https://doi.org/10.1073/pnas.1103051108

Chapter 3
OriGami: A Mobile Geogame for Spatial Literacy

**Thomas Bartoschek, Angela Schwering, Rui Li,
Stefan Münzer, and Vânia Carlos**

3.1 Introduction

Spatial literacy, the skill of learning about and improving interaction with one's surroundings, is an inherently transdisciplinary competency transcending from STEM to social sciences and arts. Spatial literacy is central in primary and secondary school curricula in many countries, and not only possesses the potentials of individual success but also fosters the importance of spatial information use in society. There is wide agreement on the transdisciplinary power of spatial thinking: Goodchild (2006) pointed out its importance for curricula in all subjects: from STEM[1], to social sciences and arts. Many tasks in the most recent PISA study (OECD 2014) on general problem-solving refer to spatial problems. The National Research Council (NRC) report "Learning to think spatially" suggests solutions for geographic information systems (GIS) as a support system to think spatially (Committee on Support for Thinking Spatially 2006). Approaches using minimal GIS for all grade levels at school, when particular spatial concepts were used incidentally, follow this direction in several studies (Bartoschek et al. 2010; Battersby et al. 2006; Marsh et al. 2007). Curricula all over the world reflect spatial competency

---

[1] Short for science, technology, engineering, and mathematics.

T. Bartoschek (✉) • A. Schwering
Institute for Geoinformatics, University of Muenster, Muenster, Germany
e-mail: bartoschek@uni-muenster.de

R. Li
University at Albany-SUNY, Albany, NY, USA

S. Münzer
University of Mannheim, Mannheim, Germany

V. Carlos
University of Aveiro, Aveiro, Portugal

© Springer International Publishing Switzerland 2018   37
O. Ahlqvist, C. Schlieder (eds.), *Geogames and Geoplay*, Advances in
Geographic Information Science, https://doi.org/10.1007/978-3-319-22774-0_3

training, although there are large differences in the way countries implemented this; the spatial tasks, level of abstraction, and learning stage of lessons in spatial literacy differ substantially.

Technologies such as GPS, tagging technologies, and sensors on smartphones have become widely available at reasonable cost and young people are very eager to use them. Despite their omnipresence, they are still insufficiently integrated into current teaching and learning practices. Spatial literacy is mainly taught in paper and pencil tasks. This is also due to the lack of suitable educational geogames that provide out-of-the box solutions for teachers. Current geogames often lack an educational concept which limits their use in schools.

Meanwhile, as the ease of access to mobile devices by students increases in many educational settings, the debate around concepts such as Bring Your Own Device—BYOD (Attewell 2015) and Mobile Learning (Clarke and Svanaes 2015; Naismith et al. 2004; Sharples 2006), and their educational potential, gain acuity. Digital games are one of the emerging educational resources which allow students to develop social skills such as teamwork and simultaneously gain experience in the use of digital technologies, in addition to learning about specific content (Prensky 2007). The gamification elements of a given technology should improve enjoyment and engagement for the user (Fudenberg and Levine 1998; Deterding et al. 2011a).

Baker et al. (2015) report that the knowledge of the educational potential of geotechnologies remains scarce and inconsistent in the field, lacking well-designed, systematic, multidisciplinary and replicable studies despite profuse mentions in the literature regarding the educational potential of geotechnologies. Accordingly, the authors propose a research agenda around four pillars: relations between the geotechnology and spatial thinking; learning geotechnology; curriculum and student learning using geotechnologies; and teacher's professional development in geotechnologies (Baker et al. 2015).

This research aims to close this gap and develop a game which is not based on existing GIS. Rather, we adopt an interdisciplinary perspective to support spatial thinking by fostering skills for orientation, wayfinding, and map comprehension. We integrate these skills in a game that adapts concepts of game-based learning to make the learning experience more enjoyable and engaging (Deterding et al. 2011b). Educational geogames—location-based games making use of mobile and geotechnologies that train spatial literacy through spatial tasks—are entertaining and support the development of spatial skills and competences (Schlieder 2014). This is especially true for mobile geogames, which are based on the movement of the player through real environments. These games have an impact on the user's perception of his or her environment and support the development of spatial competencies (Schwering et al. 2014).

In this chapter we present our app **OriGami** (Orientation Gaming), which is a game to support users to enhance their map comprehension, orientation, and wayfinding skills. According to developmental stage theories (Newcombe and Huttenlocher 2000; Piaget and Inhelder 1975), and the description of spatial competency development in most curricula (German Association for Geography (DGFG) 2007; Republic of Rwanda MoE 2006), our target user group is children ages 8–12, and young adults. The app developed within this project follows the curricular requirements ("spatial orientation") and practical requirements, since schools do not concentrate on outdoor activities for classes of geography (Hemmer et al. 2007, p. 74) where spatial competencies can be fostered. We tested the underlying concept of **OriGami** in a study, where a map-based route-following task (a variant of the game) was examined with respect to the spatial perspective of verbal route instructions. Moreover, we investigated the relation between game performance and spatial abilities. Based on these findings, we develop a complete concept of the educational game **OriGami**.

In the remaining sections of the chapter, we give an overview of spatial competencies in school curricula from different countries, from which we develop curricular requirements for geogames and discuss how existing geogames meet these requirements (Sect. 3.2). We introduce our prototypical implementation of **OriGami** and its educational concept (Sect. 3.3) and present an empirical study evaluating **OriGami** as educational geogame (Sect. 3.4). The final part of the chapter outlines the implications from the empirical study for the development of geogames and describes the complete concept of **OriGami** (Sect. 3.5).

3.2 Related Work: Spatial Ability, School Curricula and Geogames

3.2.1 Individual Differences in Spatial Abilities

Spatial abilities relate to a person's cognitive, perceptual, and information-processing capacity that characterizes individual differences in performance involving spatial information (Allen et al. 2004). Researchers such as Linn and Petersen (1985) have suggested three major categories of specific abilities that compose spatial abilities comprehensively. *Spatial visualization* describes having skill in solving spatial tasks that involve multiple steps using both visual and verbal strategies. *Mental rotation* refers to skill in imagining how a figure or object would look when rotated in two- or three-dimensional space and *spatial perception*, which is representing an object's orientation in the appropriate frame of reference despite competing perspectives or reference frames. In these following paragraphs, we explain how a specific category of spatial abilities is related to the spatial literacy that we address.

Spatial visualization, defined as the ability to store and manipulate mental visual-spatial representations (see Steck and Mallot (2000) for a review)—plays an important role for learning from external visualizations. Visual-spatial abilities were found to be an important predictor of spatial configurational learning in previous studies if the to-be-learned environment was actively or passively studied from visual media (OECD 2014; Hegarty et al. 2006; Waller 2000). Walking an unknown route through a real, unknown building was related to the ability to encode visual-spatial information as measured with the hidden patterns test

(Münzer and Stahl 2011). The route was studied from different visualizations shown on a tablet computer (maps, pictures of decision points, animation of the route through a virtual building) at the entrance of the building. It is thus expected that reading a map for following a route would be associated with spatial abilities, i.e., participants with lower abilities would make more errors in a route-following task.

Perspective taking and mental rotation. Following a route on a map may involve corrections for misalignment if route instructions utilize the egocentric perspective (e.g. "on the next intersection, go left" when the current orientation following the route is to the south). Thus, following a route on a map may require shifts of spatial perspective from the allocentric into the egocentric perspective. Perspective taking is a particular spatial mental transformation that can be measured (Couclelis 1996; Burigat et al. 2006). It might be expected that participants with lower perspective taking ability will make more errors in the route-following task. Alternatively, mental rotation ability might play a critical role. Mental rotation is a mental spatial process that has been described as "analogous" because the mental process seems to resemble an overt rotation of an object in space. The mental rotation process can best be demonstrated with chronometric measurement in which reaction times are related to disparity angles (Shepard and Metzler 1971). It has already been shown that the alignment effect when learning schematic (simple) maps is dependent on mental rotation ability (Schlieder 2014). However, the alignment effect and its relation to spatial abilities have not been studied yet in the context of following a route while reading a naturalistic, ecologically valid map.

Encoding. Individuals may differ in their ability to encode spatial information from a visualization that is shown on a computer screen. A map can be considered a complex visualization. Verbal route instructions require search processes on the map that may focus on particular aspects while ignoring others. A test that measures the ability to encode a spatial figure and recognize this figure in a more complex visual pattern may capture this requirement.

3.2.2 Spatial Competencies in School Curricula

Spatial competencies are part of (probably) all school curricula in the world. Table 3.1 reviews curricula in four different countries and lists the spatial competencies and/or tasks that students have to perform at different ages to demonstrate comprehension of basic concepts of cartography and map use. Tasks to train spatial literacy include localization of common places (e.g. home or school) and areas (e.g. neighborhoods or countries) at different scale and orientation in real space. Most tasks concentrate on maps, such as teaching map-making by sketching a map or drawing the route on a map and understanding cartographic principles. While the tasks are similar—typical tasks are to locate different places such as home, school, town, country on a map, navigate to/from school, represent different objects of the surrounding on maps, and tasks reflecting basic cartographic principles such as scale and orientation—the age at which children are supposed to study the spatial

Table 3.1 Comparison of spatial competences in three different curricula/educational standards (references for Rwanda: Ministry of Education RoR (2006), Germany (NRW): Nordrhein-Westfalen (2008), Portugal: Ministério da Educação DdEB (2001), Brazil: Ministério da Educação e do Desporto SdEF (1998))

	Rwanda	Germany (NRW)	Portugal	Brazil
1 + 2 Grade	• Locate home + village • Describe main components of way to school. • Draw route from home to school • Locate school • Identify main components of school environment	• Explore way to school and school environment • Explore important facilities in hometown • Orientate along points of interest and traffic signs • Draw way to school	[Environmental studies] • Identify basic elements of the surroundings • Represent own home • Represent own school • Represent own itineraries • Locate relatively to a reference point • Trace own route on the neighborhood plant	• Concepts of scale and their importance for spatial analyses in geography • Use of cardinal points, referencing in maps; • Cartographic orientation • Geographical coordinates • Use of maps for Orientation in daily life • Localization and representation in maps
3 + 4 Grade	• Locate sector on map of the province • Draw simple sketch map to show district and its neighboring districts.	• Use maps and other media (compass, sun) for orientation • Explore and describe regional structures (agriculture, rural areas, cities, industry) • Explore and describe changes in geographic space on all scales	• Locate at the Map of Portugal • Locate on the map the towns in the district • Map the country's capital. • Locate on the map the district capitals	• Localization and representation of positions in the class room, at home, in the city • Creating, organizing and reading of map legends
5 + 6 Grade	• Locate the Province on the map of Rwanda. • Use longitudes and latitudes to locate Rwanda on the map of East Africa • Draw a map highlighting the features identified.	• Basic orientation knowledge at different scales • Familiar with basic raster and orientation systems • Determine location in real space with aid of map + other aids for orientation • Move in real space with the aid of maps + other aids for orientation	[History and Geography of Portugal] • Identify different forms of representation • Define map • Interpretate maps • Interpret the concept of scale through observation and comparison of maps with different scales. • Use the directions of the compass for guidance	• Analysis of thematic maps on city, state and Brazil level • Analysis of different kinds of maps (topographic, touristic, climate, vegetation…) • Sketching simple maps for analyzing information and realizing correlation between facts

concepts differ. The educational standards show growing levels of abstract spatial competences, from understanding children´s daily lives and surroundings, to absolute and relative location, and further to the representation of phenomena.

3.2.3 Curricular Requirements for Geogames

After reviewing primary and secondary school curricula and educational guidelines in relation to spatial competencies in various countries, we identified the following curricular requirements with respect to spatial literacy for educational games.

Requirements with respect to spatial literacy. Orientation and map comprehension are the most central competencies mentioned in all reviewed curricula. Thus, educational geogames need to train these competencies. An educational geogame should include the following components with respect to orientation and map comprehension:

Orientation:

- Localizing oneself
- Relating the map to the real world
- Aligning the map with the real world or mental rotation

Map comprehension:

- Understanding cartographic basic principles such as abstraction through categories, symbols/legend, scale, or coordinate systems)
- Understanding coordinate systems and the concept of map scale
- Describing spatial characteristics of the environment

Tasks with different degrees of difficulty can reflect the children's differential learning stages in different grades. Mobile, digital visualizations allow for adapting the degree of difficulty, for example using the compass heading to auto-rotate maps to facilitate the task of orientation or using GPS coordinates to automatically visualize a player's location to facilitate localization Another example might be changing the symbols on a map from abstract to example-based photo-realistic 3D objects to facilitate the understanding of map categories. Being able to adapt games to the player's learning stage has an effect on how spatial tasks are realized, thus it is a requirement from a spatial literacy perspective, and also from a game perspective.

Requirements with respect to game based aspects. A successful educational game motivates students while training on spatial competencies. Digital game-based-learning can be understood as "a marriage of educational content and computer games" (Prensky 2007). From the digital game-based-learning community, we identified several requirements that educational games have to account for:

Game Elements

- Teamwork—How many players are supported; can their actions be coordinated
- Competition—Players either compete against each other or build teams to solve a joint task

- Game Type—Type of game e.g. the goal of the game, temporal properties such as real-time, and spatial properties such as the size of the game area
- Adaptability/Customization—Adapting the task to different complexity levels is an important aspect in educational games to adjust the game to the individual learning stage of the player

In schools, spatial competencies are trained with paper maps. Using geospatial technologies allows us to incorporate different didactic concepts:
Technological aspects

- Mobility—Being mobile should be reflected in the overall goal and logic of the game.
- Real Time/Environment—Real time and real environment games enable pupils to experience and comprehend orientation and map comprehension tasks in more realistic situations than in the class room, thereby increasing student motivation.
- Positioning—Positioning techniques allow us to localize the user. Games can make use of positioning techniques to support user localization by giving hints or feedback and to develop an interesting game design.
- Mobile Device Compatible—Mobile devices are a practical pre-requisite for the game being played in real environments, thus the game must be mobile device compatible. In the mobile context, low energy consumption is beneficial.

3.3 Overview and Analysis of Geogames

Mobile geogames are based on orientation and movement of players in a geographic environment. This section introduces popular mobile geogames and reviews them with respect to the requirements and their relation to spatial competency training. We first give a general introduction of each game and then summarize their characteristics.

Ingress[2] is a massive multiplayer game based on an augmented reality map, which shows a player's own position. Players have to choose a faction from two, and control augmented "portals" by physically moving to their virtual position in the real world. While it meets the majority of the gaming requirements (except the access to customization), it lacks some educational concepts of supporting spatial literacy such as systematic tasks to align the map with the real environment and competency of understanding map concepts such as changing scales and map symbols.

Actionbound[3] is an interactive geogame-app for mobile devices based on the classical geocaching principle of using GPS coordinates to find a place or item of interest. It displays the position of the gamers' devices on the base map (google

---

[2] https://www.ingress.com/.
[3] https://en.actionbound.com/.

maps) using GPS and motivates users to playfully discover the environment by accomplishing tasks related to history, politics, and culture by taking pictures. Regarding critical aspects of supporting spatial literacy, some major aspects such as adapting the degree of difficulty in aligning maps with real environment, understanding map concepts such as map symbols are missing.

MapAttack[4] is an open-source multi-player geofencing game. Geofencing games create virtual barriers or 'fences' for players to keep within or outside of. The game's goal is to rapidly collect virtual points on a map comparable to the old arcade game Pac-Man. This game has its merits of training teamwork and spatial strategic relating to how to finish the game quickly. However, it does not adequately support spatial competency in map comprehension and adaptation to individual learning stages regarding orientation and map reading.

GeoTicTacToe[5] (Schlieder et al. 2006) allows two players or teams of players to compete. Each player tries to outperform the opponent in a mapping contest by being the first to contribute a piece of information about a geographic location. As a game specially designed for map reading, finding the geographic location under time-pressure increases competition and trains fast map reading. It addressed majority of aspects in training spatial competency while missing only map orientation and map scale in the game.

City Poker[5] (Schlieder 2005) is a real-time game with the aim to get the best cards in a round. Cards are hidden as geocaches in a game area, displayed on mobile devices. A player's position is acquired through the device's GPS receiver and thus orientation in the real environment is facilitated. The players search for the caches to change cards at hand. Hints and multiple choice quizzes allow a better location precision of caches. Each team knows which cards the other team possesses as well as which cards are hidden which motivates teams to compete. While acknowledging its merits in gaming design and motivational aspects through competition, we did not find strong support in spatial literacy training such as map orientation and map comprehension.

Neocartographer[5] is a geogame project for high school students with a main objective to understand the spatial decision of gamers (Feulner and Kremer 2014). It combines learning content and real presentation on game board. Teams conquer areas by occupying and solving spatial tasks to extend their background knowledge about a geographic location. The game board is based on a street map with an overlay of virtual areas, showing also the occupied areas of the opponents. This game supports spatial literacy in some ways. For example, a good strategy and map reading as well as communication are necessary to win the game. Some aspects such as map orientation and map comprehension are not reflected in this game.

Feeding Yoshi (Bell et al. 2006) lets players collect fruits (i.e. virtual points) to feed to the character Yoshi. Fruits are displayed on a map, and are representations of nearby wireless networks. Regarding the inclusion of gaming concepts, one consid-

---

[4] mapattack.org.

[5] http://www.geogames-team.org/.

eration is that in obtaining food for Yoshi, players will not walk their usual daily route. Additionally, since there is no competition designed in this game, players may get bored after a while. However, it does support spatial literacy through individual training of orientation competency and map comprehension.

Table 3.2, lists the specific review of each games corresponding the requirements of spatial literacy and gaming design.

In general, most games will meet the requirements in the gaming aspects but lack systematic consideration of aspects that support spatial literacy. For instance, most games are based on the collection of points and other scoring mechanisms. They are instantly playable, allow for multiple players or teams and are being played in real time. The games' purposes range between data collection, leisure and education. Maps are the core elements in all games, mostly street maps at a city/neighborhood level. Regarding supporting orientation competency, most games, except Feeding Yoshi, have enabled a localizing module so that players are aware of their current location. Players can relate themselves to other objects on the map such as streets, crossings, buildings to establish spatial orientation. All games, however, do not consider training explicitly a player's map orientation competency and map comprehension. Map orientation trains a player to utilize mental rotation to align a map in games with the current environment while completing a game. Map comprehension includes skills of understanding map elements such as scale and symbols. However, in most cases, maps are just used as a platform for games instead of as an educational concept. Consequently, we do not find that typical spatial competencies taught in schools are well-supported by these existing games.

3.4 Educational Concepts for Training and Measuring Spatial Literacy

3.4.1 The OriGami Prototype

To study the educational concept of map reading, we implemented a prototype of the **OriGami** game. We implemented different navigation tasks where people follow a route on a map guided by verbal instructions. The task is intended to foster orientation competence with maps. The implementation is based on the ESRI Java Script API and IONIC Framework mobile app development. The game is implemented as an app for browsers or tablets. Depending on the platform it can be used in mobile condition with GPS or in stationary condition. It consists of a simple base map and displayed route instructions of varying complexity, for example using egocentric directions, cardinal directions, landmarks and distances. The instructions can be provided and edited by the teacher or the game leader in an online editor.

Feedback, hints and game elements allow the user to orientate and find reference points in the map and in the real world. A blue circle in the map indicates the current

Table 3.2 Overview of mobile geogame elements and classification according to spatial literacy and game-based aspects

	Ingress	Actionbound	MapAttack	GeoTicTacToe	CityPoker	Feeding Yoshi	Neocartographer
Spatial Literacy Aspects: Orientation and Map Comprehension							
Localizing yourself	X	X	X	X	X	X	X
Orientation aids	X	X	X	X	X	X	X
Mental rotation/alignment	–	–	–	–	–	–	–
Type of map /cartographic elements	Streets visualized as lines	Satellite, Street map, topographic	Street map	Street map	Street map	Street map	Street map
Alternative map type	–	X	X	–	–	–	–
Virtual elements/augmentation	Portals, links	–	Points to be collected	9 boxes	Five regions	Fruits	Game area
Auto rotation	X	–	X	–	–	–	–
Game-based Aspects (Game Elements and Technological Aspects)							
Single/multiplayer	Multi-player	Both	Multiplayer	2 players	2 players/ teams	Multi-player	Multi-player
Coordination of actions	X	X	Yes	–	–	Chat	X
Classical video game elements	Score system	Score system	Score system	–	–	–	–
Competition (speed/points)	Speed, points	–	Speed, points	Speed, time	Speed/Best cards	Points	speed, time, points
Main focus	strategy	Learn environment	Speed	Strategy	Speed/strategy		Learn environment
Game time/duration	Unlimited/ongoing	Variable	short	Short	Variable	Unlimited	Ca. 45 min
Real time or turn based	Real-time	Real-time	Real-time	Real-time	Real-time	Real-time	Real-time
Size of game area	global	Variable, city level	Variable, street level	Variable	Variable (city level)	City-level	Variable, city-level
Adaptability/customization	–	Create own bounds	Create game board	–	Game editor	–	ArcGIS Online
Change/add elements by player	X (portals)	–	–	–	–	–	–

position located by the device or selected by the user by clicking on the map. A smiley face provides feedback on the current walking direction: It changes color and friendliness (smile or scowl) to give intuitive hints about whether the player is moving or clicking in the direction of the next waypoint. Each time the player reaches a waypoint (in the browser version there is a tolerance distance for clicking, depending on the zoom level), the app signals this by playing a sound and visually via a happy smiley and a text message. The next wayfinding instructions are automatically displayed at the bottom of the screen. A trumpet sound and a text at the end of a route give users feedback, that they have reached the goal. The interface design is kept extremely simple, choosing the map as the main element covering the screen (Fig. 3.1).

OriGami is a game but at the same time a measurement tool to evaluate the spatial literacy and learning progress of a player. The app has several possibilities to record the user interactions for analysis in terms of usability or learning. The tablet version records the GPS-Track, the touch coordinates and gestures, the zoom level of the map and the time required for each route. For optional thinking-aloud tests it can also record sound and allow further usability analysis. The browser version records the time from loading the route to successfully finishing it by reaching the goal, each click coordinate, the distance to the actual waypoint and the zoom level of the map at each click. These designs allow the use of the app and its recorded data as a variable in tests on spatial learning.

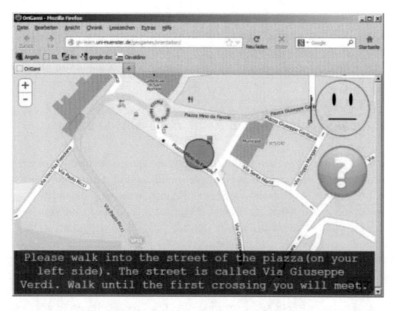

Fig. 3.1 Screenshot of the **OriGami** App in the browser version

3.4.2   Evaluating the Educational Concept behind OriGami

The goal of the **OriGami** game is to train spatial competencies. For this study we selected different aided navigation tasks and tested the performance of students in different conditions of map-based route-following tasks with the browser-based **OriGami** prototype in a lab environment. As explained below, we assume that spatial abilities have an effect on the performance and the effect depends on the frame of reference used for the route instructions.

Our target user group is children at the age of 8–12 as well as young adults in accordance to the developmental stage theories (Newcombe and Huttenlocher 2000; Piaget and Inhelder 1975) and the description of spatial competency development in most curricula (German Association for Geography (DGFG) 2007; Republic of Rwanda MoE 2006). The app follows the aforementioned curricular requirements of teaching spatial literacy and game-based requirements, and since schools usually use indoor activities for geography lessons (Hemmer et al. 2007, p. 74) we provide the browser-based version.

Before conducting a long-term study on training effects, we evaluate (1) whether different conditions of the game reasonably impose different levels of difficulty, (2) whether individual differences of spatial ability are related to initial performance, and (3) whether different levels of map reading expertise affect performance. The present study pursues these goals. Studies like these help us to explain different performances and interpret learning progress of individuals for different ways to communicate about space with different reference frames (allocentric, egocentric, landmark-based). Examples of these three types of instructions are shown in Table 3.3. They also help to understand which user groups **OriGami** should address. This study does not measure the effect **OriGami** has on spatial literacy. A study measuring the learning progress while playing **OriGami** would require a long-term study and a procedure to measure spatial competencies, which is still future work.

3.4.3   Spatial Perspective in Route Following on a Map

The level of difficulty of the game varies with the spatial perspective of the verbal route instructions, because the perspective might require particular cognitive processes. The present study investigated performance in the map-based learning game in different route instruction conditions. Participants were asked to follow a route by clicking waypoints that were verbally described. Route instruction conditions differed mainly with respect to the spatial perspective of route instructions adopted for the game (allocentric route instructions, egocentric route instructions, and landmark-based route instructions). Route instructions using the allocentric perspective directed players to an object location based upon the position of other players. In this type of route instruction, the cardinal directions e.g. south, north were used but a person's location was not used. Route instructions given using the egocentric

Table 3.3 Examples (translated from German) for an initial instruction at the start point and for an instruction at a waypoint in all three conditions (landmark, egocentric, allocentric)

	Landmark	Egocentric	Allocentric
Initial instruction for orientation	On your right you see the Sacred Heart Church. On your left you see the Old Postman Pub. Go to the next junction.	You are looking in the direction of "Cologne Street". Go straight until you reach the next junction.	Go south until you reach the next junction.
Instruction at a waypoint	Turn and go to the Art house.	Turn right and walk along the street until you reach the second junction	Turn north and walk until you reach the second junction.

perspective directed players to the destination object with respect to its relative direction to a person's standing position. The route utilized relative direction such as left or right in this type of instruction. In the land-mark-based route instructions, all instructions were created through the interrelation between landmarks. Terms such as "toward" were used to indicate the direction.

Since maps are typically studied with the particular north-on-top orientation, judgments of relative directions from memory are more difficult if they require reorientation, e.g., imagining another position and orientation than the orientation from which the representation has been studied. Mental representations of spatial configurations are thus thought to be orientation dependent (Schwering et al. 2014; Baudisch and Rosenholtz 2003; Flügel 2014; Bitgood 1991; Robinson 1928). The orientation specificity effect is also termed alignment effect. Alignment effects are considered robust (Schwering et al. 2014; Feulner and Kremer 2014). They occur both with large and small layouts (Flügel 2014). The effect can be experienced in everyday spatial activities such as navigation. For instance, misaligned you-are-here maps impede orientation in a real environment (Deterding et al. 2011b; Li et al. 2014; Gunzelmann and Anderson 2006). This suggests that alignment plays a role in map reading when planning a route and not only for retrieval from memory. In the present study, naturalistic maps were utilized in different route instruction conditions. It was expected that an instruction that describes the route from an ego-centric point of view would cause alignment problems when participants try to follow the route.

3.5   Empirical Study on the Educational Concept

The goal of the game is fostering map reading competences. A precondition of this goal is to establish a difference in performance between experts and non-experts in map reading when they play the game initially. Experts are those who are familiar with geospatial concepts such as location, distance, or direction and who have received training in map reading, map projection, or coordinate systems in curriculum, while non-experts are those who have not received those trainings in their curricula. This difference would show that the game reflect differences in expertise with its tasks. Therefore, the present study compares two groups of participants. The expert group comprises participants who study geoinformatics, computer science, and landscape ecology. These participants are expected to have more experience with spatial processing, more prior knowledge about maps and higher map reading skills than participants in the non-expert group. The non-expert group comprises participants who study education and teaching, mathematics, psychology, or history. It is expected that the experts group will outperform the non-experts group in the route following tasks of the game, particularly in the most difficult condition.

Additionally, map reading to follow the route requires search processes and corrections for alignment, particularly in the egocentric route condition. These processes may depend on spatial abilities and on acquired competencies. The ego-centric route condition was therefore expected to be more difficult because mental processes of perspective taking were inevitable.

3.5.1 Methodology

Participants. Forty-eight participants took part in the experiment, 26 of whom were female (n = 48). They were students at the university of Münster or at the University of Mannheim, Germany, and studied education and teaching (n = 20), geoinformatics (n = 15), psychology (n = 6), landscape ecology (n = 6), computer science (n = 2), history (n = 1) and mathematics (n = 1). The average age was M = 24.3 (SD = 4.9). Participants received course credit or remuneration for participation.

Materials. Following our introduction of spatial abilities, we utilize the following psychometric tests that correspond to specific aspects of participant's spatial abilities: the hidden patterns test for spatial visualization; the perspective taking tests for spatial perception, and mental rotation test for mental rotation ability.

The hidden patterns test (Guay 1976) measures encoding and recognizing a simple figure which is embedded in a more complex line drawing (Fig. 3.2). Two-hundred items were shown on four pages. Participants answered by marking answer boxes below the items. The overall processing time was restricted to 3 min. In the scoring procedure, the number of incorrectly marked answers was subtracted from

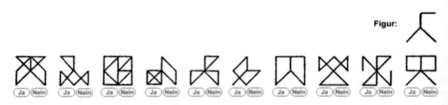

Fig. 3.2 Sample item of the hidden patterns test

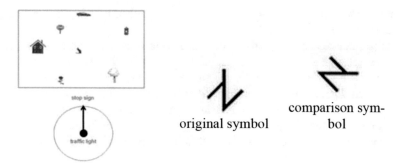

Fig. 3.3 *Left*: Perspective taking test map: The answer is marked on the *circle*. *Right*: Example symbols presented on Chronometric mental rotation test

the number of correctly marked ones. A reliability of 0.91 is reported for this test (Guay 1976, p. 11).

In the perspective taking test (Couclelis 1996; Burigat et al. 2006), participants are asked to make directional judgments based on a map which shows a spatial configuration of seven objects (Fig. 3.3). Participants imagine themselves standing at a particular position (e.g., at the traffic light), facing a particular second location (e.g., the stop sign), and pointing to another location (e.g., the flower). The directional judgment is indicated by a position to be marked on the answer circle. The map is visible during answering. The test requires estimating directions from imagined positions with orientations that deviate from the "upright/north" orientation of the map typically more than 90°. Participants process 12 items, all utilizing the same map. The score of the participant is the average angular error calculated from the items that the participant attempted to solve within the given time of 5 min. Reliability estimates between 0.79 and 0.85 (Cronbach's alpha statistic) are reported for this test (Couclelis 1996).

A computer-based test on mental rotation ability including reaction time measures was created after a description provided by Gustafson et al. (2008), using PMA symbols (Krukar and Conroy Dalton 2013). For each item, an original symbol and a comparison symbol was shown on the screen (Fig. 3.3). The comparison symbol was rotated with an angular disparity of 0°, 45°, 90°, and 180°. The comparison symbol either was identical to the symbol on the left, or it was mirrored. Participants were asked to determine as quickly as possible whether the two symbols were identical or not. The test included 60 items. Reaction times as well as accuracy (number of wrong answers) were measured. Jansen-Osmann and Heil (2007) estimated reliability with the Odd-Even method and reported r = 0.91 for the reliability of this test.

Materials: Route Instructions. We selected three routes in three different urban locations in Germany for the route-following task using the **OriGami** prototype described above. The three routes are comparable in complexity: ten instructions had to be executed to reach the goal. The routes contain eight turns at waypoints and in-between waypoints. For each route we created landmark-based, egocentric and allocentric route instructions (Table 3.3).

The landmark-based route instructions used solely landmarks to give directions. All landmarks were visible on the highest and second-highest zoom level. They were represented either via a symbol, a label, the footprint of the landmark, or a combination. Egocentric instructions used only egocentric turn directions. Only for the initial orientation at the start point a landmark was used. The allocentric instructions used only allocentric turn directions. All three route instructions used the terms junction, round-about, and footpath to refer to special features of the street network on the map.

Procedure. Participants were administered the spatial ability tests first (hidden patterns test, perspective taking test, mental rotation test). Subsequently, participants completed three route-following tasks corresponding to the three instruction conditions in the browser-based version of **OriGami**. All participants completed the route-following tasks in all of the three instruction conditions (allocentric, egocentric, and landmark-based route instruction condition). The order of the route instruction conditions was balanced across participants. The three different route instruction conditions were specific to three different actual routes based on three different city maps such that each participant received the three route instructions with three different routes. Each actual route was equally presented with a particular route instruction. As introduced in the description of the prototype, each participant would follow the instruction on the screen (allocentric, egocentric, or landmark-based depending on the condition) to click the waypoint on the screen. Each time the player reaches a waypoint (in the browser version there is a tolerance distance for clicking, depending on the zoom level), the app signals this by playing a sound and visually via a smiley and a text message to indicate its accuracy. The next wayfinding instructions are automatically displayed at the bottom of the screen. A trumpet sound and a text at the end of a route gives the user feedback, that a participant has reached the goal. Participants were tested in small groups of 2–5 participants in a multimedia laboratory with separate work spaces. The experiment lasted about 40 min.

3.5.2 Results and Discussion

Due to a technical error, the data of four participants were lost for the mental rotation test and the measurement of the landmark-based route instruction condition. These four cases were removed from the data set.

The data set was screened for outliers which were defined as values above 2.5 standard deviations from the mean (M and SD calculated with original data). In the perspective taking test, the average angular error of one participant exceeded 2.5 SD. In the mental rotation test, reaction times to correctly solved rotatable items exceeded 2.5 SD in two cases. These three values were replaced by the respective means. Furthermore, numbers of errors (false clicks, i.e. clicks not on the next waypoint on the map) were inspected in the allocentric, in the egocentric and in the landmark condition. Extremely high numbers of errors were found in two cases in the allocentric condition, in three cases in the egocentric condition, and in one case in the landmark condition. Errors represent clicks on the map that are placed outside a predefined area around the correct intersection or location. A closer inspection revealed that these participants obviously clicked on the complete route (i.e., they simulated walking along the streets by clicking "along" the streets). Moreover, this behavior could only be observed if the respective route instruction condition was the first of the three conditions accomplished by the participant. Thus, the incorrect clicks were attributed to a misunderstanding of the experimental instruction rather than to a particular difficulty. Therefore, the respective values were replaced by the mean of the respective variable. If a participant made an extremely high number of errors, then the time needed to complete the condition increased accordingly. Therefore, corresponding times were corrected (replaced by the mean of the respective variable) for the six cases in which the number of errors had been replaced.

Based on their main subject of study at university, participants were assigned to either an expert group or a non-expert group. The expert group comprised participants whose majors were geoinformatics, computer science, and landscape ecology ($n = 20$, four participants in this group were female). These participants were those who have taken geospatial curricula in the institute of geoinformatics in which basic spatial concept and map concepts were addressed. Because of their curricular trainings, participants in this group were expected to have more experience with spatial processing, more prior knowledge about maps and higher map reading skills than participants in the non-expert group. The non-expert group comprised participants whose major were education and teaching, mathematics, psychology, or history ($n = 24$, 18 participants in this group were female). They were recruited through an introductory psychology classes and have not had geospatial trainings in their curricula.

Table 3.4 shows descriptive data for the spatial abilities tests, the number of errors (false clicks) in each condition, and the time (seconds) needed to complete each condition, separated for the expert and the non-expert group. Experts outperformed non-experts in the perspective taking test, $t(42) = -3.088, p < 0.01$, as well as in the hidden patterns test, $t(42) = 3.910, p < 0.001$. However, mental rotation ability (as indicated by the time needed to assess rotatable items in the chronometric mental rotation test) did not differ between groups, $t(42) < 1$, ns.

Table 3.5 shows the correlations between the measured variables. Intercorrelations between the spatial abilities tests were in the medium range ($r = 0.38$ and $r = 0.45$). Mental rotation ability predicted the numbers of false clicks in the allocentric and in the egocentric condition significantly. Moreover, mental rotation predicted the time to complete the allocentric condition specifically. The hidden patterns test only predicted the time to complete the allocentric condition. The perspective taking test did not correlate significantly with any of the route performance measures. Within a route condition, correlations between the numbers of errors and the time to complete the route were substantial ($r = 0.61$ in the allocentric condition, $r = 0.77$ in the egocentric condition, $r = 0.55$ in the landmark condition), suggesting that more errors caused longer completion times. Between route conditions, numbers of errors in the allocentric and in the egocentric conditions were related ($r = 0.45$), but errors in the landmark condition were unrelated to the other two conditions. Times to complete the route conditions were not interrelated.

In order to analyze effects of route condition and group, two repeated-measures analyses of covariance were performed, separately for errors and times. In these analyses, route condition was included as the within-subject factor with three levels, and group was included as the between-subject factor with two levels. In addition, mental rotation (time) and hidden patterns (test score) were included as co-variates. Both covariates were centered. (Perspective taking ability was not included as a covariate, because the correlations had shown that perspective taking did not predict performances in the route tasks).

In the analyses of the errors, a Mauchly test indicated violation of sphericity (Mauchly-W = 0.80, $p < 0.05$). Therefore, the Greenhouse-Geisser correction was applied. The analysis revealed a main effect of route condition, $F(1.66,66.44) = 3.377$, $p < 0.05$, partial eta squared = 0.08. More errors were made in the egocentric route condition (adjusted mean 16.04, $SE = 2.47$) than in the allocentric route condition (adjusted mean 9.84, $SE = 1.44$), $p < 0.05$. Errors in the landmark condition were similar to the allocentric condition descriptively (adjusted mean 10.97, $SE = 1.4$), but did not differ significantly from the other two route conditions.

Furthermore, a main effect of expert group was found, $F(1,40) = 6.798$, $p < 0.05$, partial eta squared = 0.15. Adjusted means suggest that non-experts (16.42 errors, $SE = 1.9$) made about twice as much errors as experts did (8.14 errors, $SE = 2.13$). The main effect was qualified by a marginally significant interaction of route condition and expert group, $F(1.66,66.44) = 3.115$, $p = 0.06$, partial eta squared = 0.08. Descriptive data (Table 3.4) and a marginally significant post-hoc t-test suggested that the difference between experts and non-experts might be have been most substantial in the egocentric route condition, $t(42) = -2.002$, $p = 0.052$, whereas differences between groups were not significant in the allocentric and in the landmark condition.

Table 3.4 Descriptive data for spatial abilities and performance in the route following game in the expert and in the non-expert group (errors are the number of false clicks, times are provided in seconds)

	Experts ($n = 20$) M (SD)	Non-experts ($n = 24$) M (SD)
Perspective taking (average angular error)	17.09 (9.29)	32.82 (21.11)
Hidden patterns (test score)	245.70 (37.75)	195.46 (45.97)
Chronometric mental rotation (time in ms)	2747.39 (961.57)	2622.92 (711.42)
Errors in allocentric route condition	7.35 (6.51)	12.42 (11.91)
Errors in egocentric route condition	10.95 (12.86)	21.67 (20.84)
Errors in landmark route condition	8.75 (8.94)	13.00 (9.47)
Time to complete allocentric route condition	217.80 (114.33)	231.33 (104.79)
Time to complete egocentric route condition	221.40 (100.84)	286.58 (166.11)
Time to complete landmark route condition	316.50 (117.54)	293.33 (108.14)

The analysis also revealed a main effect of mental rotation ability $F(1,40) = 7.637$, $p < 0.01$, partial eta squared $= 0.16$, and a significant interaction of route condition and mental rotation ability, $F(1.66,66.44) = 5.390, p < 0.05$, partial eta squared $= 0.12$. This means that the relation between mental rotation ability and errors was not the same in each route conditions. Correlations (Table 3.5) show that mental rotation predicted errors in the allocentric route condition and in the ego-centric route condition, but not in the landmark route condition.

Results for times to complete the route following task showed a main effect of route condition on times, $F(2,80) = 5.844$, $p < 0.01$, partial eta squared $= 0.13$. However, the pattern differed from that found for errors. Participants seemed to need more time in the landmark condition specifically (adjusted mean 306.45 s, $SE = 16.83$), compared to the allocentric condition (adjusted mean 226.28, $SE = 14.87$) and to the egocentric condition (adjusted mean 252.49 s, $SE = 21.06$). However, only the allocentric condition differed significantly from the landmark condition, $p < 0.01$, whereas the egocentric condition did not differ from the other two conditions.

A main effect of expert group was not obtained (adjusted means for experts 259.27 s, $SE = 19.63$, for non-experts 264.27 s, $SE = 17.53$), however, there was a significant interaction between route condition and expert group, $F(2,80) = 3.757$, $p < 0.05$, partial eta squared $= 0.09$. Whereas single comparisons between the expert and the non-expert groups did not show significant differences in any of the route conditions, descriptive data seem to suggest that experts needed more time to complete the landmark condition than the non-experts. The reverse was true for the allocentric and the egocentric conditions. The significant interaction might correspond to this inconsistent pattern.

Spatial abilities (hidden patterns test score, mental rotation times) did not affect times to complete the route following tasks. Neither main effects nor interactions with route conditions were found.

In summary, route condition, expertise (group) and mental rotation ability affected errors (false clicks). The effect size of route condition was in the medium range, whereas effect sizes of expertise and mental rotation ability were large. Corresponding to errors, times were affected by route condition. However, times were neither affected by expert group or nor by spatial abilities. Results suggest that the landmark route condition differed in requirements remarkably from the allocentric and the egocentric route condition. First, numbers of errors suggested that the egocentric condition was the most difficult condition. However, participants needed most time in the landmark condition. (Furthermore, experts seemed to need even more time than non-experts to complete this particular condition, in contrast to the other two conditions.) Second, the pattern of correlations (Table 3.5)—corresponding to the interaction between mental rotation ability and route condition—shows that mental rotation ability did not play a role for the number of errors made in the landmark condition. It is therefore possible that cognitive processing requirements were quite specific in the landmark condition.

Table 3.5 Correlations between spatial abilities tests and performance measures in the route following game in three conditions (allocentric, egocentric, and landmark condition)

	Perspective taking	Hidden patterns	Mental rotation	Errors allocentric	Errors egocentric	Errors landmark	Time allocentric	Time egocentric
Hidden Patterns	-0.38^*							
Mental rotation	-0.01	-0.45^{**}						
Errors allocentric	0.00	-0.27	0.31^*					
Errors egocentric	0.13	-0.23	0.37^*	0.45^{**}				
Errors landmark	0.25	-0.29	0.07	0.28	0.16			
Time allocentric	0.13	-0.42^{**}	0.39^{**}	0.61^{**}	0.22	-0.01		
Time egocentric	0.10	-0.14	0.21	0.45^{**}	0.77^{**}	0.06	0.26	
Time landmark	0.13	-0.18	0.18	0.02	0.13	0.55^{**}	0.14	0.09

Note. $^*p < 0.05$, $^{**}p < 0.01$. Errors are numbers of false clicks in the respective route condition. Time means the duration in seconds to complete the route following task in the respective condition

The study shows a successful evaluation of the route-following task in **OriGami**. The task (1) varies in difficulty due to particular spatial processes, such as correction of misalignment, (2) correlates with spatial ability (mental rotation) and (3) demonstrates performance differences which plausibly depended on expertise. Build upon these results, the **OriGami** game can be extended to train more spatial competencies. The study also brought more detailed questions regarding the cognitive processes associated with the tasks which can be systematically explored in an extended version of **OriGami**. In particular, the cognitive requirements of the landmark condition seemed to differ from the egocentric and the allocentric condition. Moreover, individual differences in spatial encoding and perspective taking did not play important roles in accomplishing the tasks.

3.6   Spatial Literacy Training with OriGami

Based on the study results with our first prototype presented above and based on the requirements we draw from curricula, we developed a comprehensive concept of the **OriGami** game that supports the acquisition of better spatial competencies while playing the game through a series of navigation and orientation tasks. Like in the prototype version, the player is equipped with a smartphone or tablet (which provides positioning technologies). The goal is to navigate to a certain location where you have to solve a task. Those routes are created beforehand by a game master /teacher in the inbuilt editor. The following features have being included in the final conceptualization of the game.

3.6.1   Navigation

Two different navigation types can be distinguished: an aided navigation task or a path planning task. In the aided navigation task, the player receives route instructions to the next waypoint. Instructions are given either allocentric, egocentric or landmark-based. Based on the given route instruction the player has to move in the real environment and find the next waypoint to receive the next instructions which successively lead her to the destination. In the path planning task the player receives a map of the environment and the destination visualized on this map. The player has to locate him or herself on the map and determine the best route to the destination.

Once arrived at the destination, the user has to solve a task. This task can be defined by the teacher. In principle these tasks can relate to any subject from STEM to social sciences and arts. In the following sections, we describe tasks training orientation and map comprehension. Students solve one task at each destination, but the teacher can define different task for each destination.

3.6.2   Orientation Task

The user has to georeference a photo (in more detail: the position from where the photo was taken) and add spatial or thematic data to the map that can be derived from the photo. Depending on the subject and the degree of difficulty, the photo might show different things, e.g. in case of historical photos, the scenery might have changed. New houses might occlude houses that are visible on the historic photo. The photo might show underground supply circuits that are not visible either. This way we can create tasks with different levels of complexity challenging the player's ability to read and interpret a map.

3.6.3   Map Comprehension Task: Cartographic Basics

To make abstract concepts such as coordinate systems more concrete, teachers design tasks that help to experience coordinate systems in practice. For example, students are asked to walk along longitudes, latitudes, certain degrees or angles, or walk to the most northern/southern point of the destination region (e.g. a school ground, a park, a square).

3.6.4   Map Comprehension Task: Spatial Learning

The task to draw a map from home to school is classical in most curricula. Thus, the same task can be integrated in **OriGami**. Students are asked to draw a sketch map of the route they travelled, take a picture of the sketch map, and upload it to the system. Other players of the game travelling the same route can rate previously created maps. The teacher can use this material also in an after-game class to analyze typical cartographic concepts.

3.6.5   Spatial Competency Testing and Training

The navigation task fosters different spatial competencies. Depending on the configuration of the app, we visualize a player's own location (determined via GPS) with differing degrees of precision or we do not visualize it. This way players have to determine the own position on a map. The app can be used for map alignment tasks, if the automatic alignment to users' movement direction is switched off. Cartographic basic principles such as map scale are trained by navigational aids referring to objects visible only at certain scales (e.g. local landmarks along the route or global landmarks off the route), which requires the player to change the scale of the map.

The georeferencing and mapping task also fosters spatial competencies: In the photo georeferencing task, the player has to relate objects in the map to objects in the reality. She has to align the map or apply mental rotation with different degrees of complexity (occlusion of objects, 2D and 3D rotations). The data collection task trains the player in understanding symbols and cartographic elements in maps. Different degrees of complexity range from collecting data falling in an existing thematic category to collecting data that forces the player to create new categories. Here, new thematic categories with suitable symbols and visualization need to be identified. In contrast to many other games, this app supports different map types that involve not only street networks as background map but also other map types (different providers like OSM, topographic maps or satellite images). Different map types shall point children to the advantages and disadvantages of different maps for certain tasks.

3.6.6   Game-Based Aspects

The game can be played in different modes depending on the settings that are chosen by the teacher.

Environment: The game can be played in the real environment (which ensures real experiences for the player) or in a stationary class-room mode.

Adaptability/customization: The routes in the navigation task and the task at the destination are set-up by the teacher. Furthermore, the teacher can choose different settings for the game that lead to different complexity levels when playing the game. This way the teacher can adapt the game to the learning stage of the students. An editor is provided to define routes and tasks (e.g. the upload of photos). The route instructions can be provided and edited by the teacher or the game leader.

Teamwork: **OriGami** can be played as single-user or multi-user game. In the multi-user setting, we aim for two different modes: In the collaborative mode, we focus on team aspects. It incorporates the same orientation tasks as in the basic game but adds collaborative and competitive elements where one player is the editor describing routes and the other is the scout following the instructions. This mode was inspired by game shows where a game master guides the players/teams through competitions.

Game Element: The user receives instant feedback via a smiley for his or her actions. Feedback and hints allow the user to orientate and find reference points in the map and in the real world. In general playing digital games develops soft skills of the user. Students are expected to be better at working in teams and gain experience in using geospatial technologies. ICT (Information and Communication Technologies) skills are practiced playing digital games.

3.7   Conclusion

In this chapter we review curricula specifications regarding spatial competency and spatial literacy training in an educational context. We evaluate the state of art for geogames that focus on training spatial competencies and identified a lack of tools fostering spatial competency and spatial literacy for children and young adults. Many spatial competencies are studied theoretically in school. Geogames allow users to experience many of these theoretical concepts in the real world, e.g. experience map alignment and orientation in a goal-directed wayfinding task or experience the concept of a coordinate system and cardinal directions in the real world.

The geospatial technologies required for such games exist and are robust enough to be used in educational games. Therefore, we propose **OriGami**, an educational game fostering spatial literacy. **OriGami** allows users to train specific spatial competencies through a set of tasks and measure performance in these tasks. By relating it to the user's spatial abilities, it allows teachers to individually select training tasks for specific competencies. In the empirical study, we showed the relation of spatial abilities and performance to support the need for comprehensive, curriculum-based geogames.

Future work addresses the further development of the **OriGami** game according to the game concept above. Afterwards, we intend to conduct studies where spatial literacy and competency development is measured over a longer term period to show the effect of **OriGami**.

Acknowledgments   This work was supported by the Erasmus+ grant VG-SPS-NW-14-000714-3. Furthermore, we would like to thank Matthias Pfeil for the software development and ESRI Inc. for their financial support.

References

Allen G, Kirasic K, Rashotte M, Haun D (2004) Aging and path integration skill: kinesthetic and vestibular contributions to wayfinding. Percept Psychophys 66(1):170–179

Attewell J (2015) BYOD bring your own device: a guide for school leaders. European Schoolnet, Brussels, Belgium

Baker TR, Battersby S, Bednarz SW, Bodzin AM, Kolvoord B, Moore S, Uttal D (2015) A research agenda for geospatial technologies and learning. J Geogr 114(3):118–130

Bartoschek T, Bredel H, Forster M (2010) GeospatialLearning@PrimarySchool: a minimal GIS approach. In: Sixth international conference on geographic information science, extended abstracts. University of Zürich, Zürich, Switzerland

Battersby SE, Golledge R, Marsh M (2006) Incidental learning of geospatial concepts across grade levels: map overlay. J Geogr 102:231–233

Baudisch P, Rosenholtz R (2003) Halo: a technique for visualizing off-screen objects. In: SIGCHI conference on human factors in computing systems. ACM, New York

Bell M, Chalmers DJ, Barkhuus L, Hall M, Sherwood S, Tennet P, Brown B (2006) Interweaving mobile games with everyday life. In: CHI—SIGCHI conference on human factors in computing systems. Montréal, Canada. ACM, New York, pp 417–426

Bitgood S (1991) Common beliefs about visitors: do we really understand our visitors? Visitor Behav 6(1):6

Burigat S, Chittaro L, Gabrielli S (2006) Visualizing locations of off-screen objects on mobile devices: a comparative evaluation of three approaches. In: 8th conference on human-computer interaction with mobile devices and services. ACM, New York

Clarke B, Svanaes S (2015) Updated review of the global use of mobile technology in education. Techknowledge for Schools, London

Committee on Support for Thinking Spatially, The Incorporation of Geographic Information Science Across the K-12 Curriculum, National Research Council (2006) Learning to think spatially. National Academies Press, Washington, DC

Couclelis H (1996) Verbal directions for way-finding: space, cognition, and language. The construction of cognitive maps. Springer, Berlin

Deterding S, Dixon D, Khaled R, Nacke LE (2011a) From game design elements to gamefulness: defining gamification. In: 15th international academic MindTrek conference: envisioning future media environments. ACM, New York

Deterding S, Dixon D, Khaled R, Nacke LE (2011b) From game design elements to gamefulness: defining gamification. In: Mindtrek 2011. ACM Press, Tampere

Feulner B, Kremer D (2014) Using geogames to foster spatial thinking. Herbert Wichmann Verlag, Heidelberg

Flügel K (2014) Einführung in die Museologie, 3rd edn. WBG Verlag, Darmstadt

Fudenberg D, Levine DK (1998) The theory of learning in games. MIT Press, London

German Association for Geography (DGFG) (2007) Educational standards in geography for the intermediate school certificate. Deutsche Gesellschaft für Geographie (DGfG), Germany

Goodchild M (2006) The fourth R? Rethinking GIS education. ArcNews 28(3):5–7

Guay R (1976) Purdue spatial visualization test. Purdue University, West Lafayette, IN

Gunzelmann G, Anderson JR (2006) Location matters: why target location impacts performance in orientation tasks. Mem Cogn 34:41–59

Gustafson S, Baudisch P, Gutwin C, Irani P (2008) Wedge: clutter-free visualization of off-screen locations. In: SIGCHI conference on human factors in computing systems. ACM, New York

Hegarty M, Montello D, Richardson AE, Ishikawa T, Lovelace KL (2006) Spatial abilities at different scales: Individual differences in aptitude-test performance and spatial-layout learning. Intelligence 34:151–176

Hemmer I, Hemmer M, Neidhardt E (2007) Räumliche Orientierung von Kindern und Jugendlichen—Ergebnisse und Defizite nationaler und internationaler Forschung. In: Geiger M, Hüttermann A (eds) Raum und Erkenntnis. Aulis Verlag Deubner, Köln

Jansen-Osmann, P., & Heil, M. (2007). Suitable stimuli to obtain (no) gender differences in the speed of cognitive processes involved in mental rotation. Brain and Cognition, 64(3), 217–227. doi:http://dx.doi.org/10.1016/j.bandc.2007.03.002

Krukar J, Conroy Dalton R (2013) Spatial predictors of eye movement in a gallery setting. In: Eye tracking for spatial research, proceedings of the first international workshop (in conjunction with COSIT 2013). Scarborough, UK

Li R, Korda A, Radke M, Schwering A (2014) Visualising distant off-screen landmarks on mobile devices to support spatial orientation. J Locat Based Serv 3:166–178

Linn D, Petersen A (1985) Emergence and characterization of sex differences in spatial ability: a meta-analysis. Child Dev 56(6):1479–1498

Marsh M, Golledge R, Battersby SE (2007) Geospatial concept understanding and recognition in G6-College students: a preliminary argument for minimal GIS. Ann Assoc Am Geogr 97(4):696–712

Ministério da Educação DdEB (2001) Currículo Nacional do Ensino Básico—Competências Essenciais

Ministério da Educação e do Desporto SdEF (1998) Parâmetros Curriculares Nacionais. Terceiro e Quarto Ciclos do ensino fundamental de geografia., Brasilia

Ministry of Education RoR (2006) National Curriculum Development Centre (Ed.) In: Social studies curriculum for the basic education programme

Münzer S, Stahl C (2011) Learning routes from visualizations for indoor wayfinding: presentation modes and individual differences. Spat Cogn Comput 11(4):281–312

Naismith L, Lonsdale P, Vavoula GN, Sharples M (2004) Mobile technologies and learning. Futurelab, Malaysia

Newcombe N, Huttenlocher J (2000) Making space: the development of spatial representation and reasoning. MIT Press, Cambridge, MA

Nordrhein-Westfalen (2008) Richtlinien und Lehrpläne für die Grundschule in Nordrhein-Westfalen: Deutsch, Sachunterricht, Mathematik, Englisch, Musik, Kunst, Sport, Evangelische Religionslehre, Katholische Religionslehre. Schule in NRW, Ritterbach, Frechen

OECD (2014) PISA 2012 results in focus: what 15-year-olds know and what they can do with what they know. OECD, Paris

Piaget J, Inhelder B (1975) Die Entwicklung des räumlichen Denkens beim Kinde Gesammelte Werke 6 (Studienausgabe), 3rd edn. Klett-Cotta/J. G. Cotta'sche Buchhandlung Nachfolger, Germany

Prensky M (2007) Digital game-based learning. Paragon House, St. Paul, MN

Republic of Rwanda MoE (2006) Social studies curriculum for the basic education programme, grades 1–6

Robinson ES (1928) The behaviour of the museum visitor. The American Association of Museums, Washington

Schlieder C (2005) Representing the meaning of spatial behavior by spatially grounded intentional systems. In: GeoSpatial Semantics. Springer, Berlin, pp 30–44

Schlieder C (2014) Geogames–Gestaltungsaufgaben und geoinformatische Lösungsansätze. Informatik-Spektrum 37(6):567–574

Schlieder C, Kiefer P, Matyas S (2006) Geogames—designing location-based games from classic board games. IEEE Intelligent Systems, Special Issue on Intelligent Technologies for Interactive Entertainment. p 40–46

Schwering A, Münzer S, Bartoschek T, Li R (2014) Gamification for spatial literacy: the use of a desktop application to foster map-based competencies. In: AGILE workshop geogames. Utrecht University, Netherlands

Sharples M (2006) big issues in mobile learning: report of a workshop by the kaleidoscope network of excellence—mobile learning initiative. The University of Nottingham, UK

Shepard RN, Metzler J (1971) Mental roation of three-dimensional objects. Science 1(71):701–703

Steck SD, Mallot HA (2000) The role of global and local landmarks in virtual environment navigation. Presence 9:69–83

Waller D (2000) Individual differences in spatial learning from computer-simulated environments. J Exp Psychol 6(4):307–321

Chapter 4
Spatial Game for Negotiations and Consensus Building in Urban Planning: YouPlaceIt!

Alenka Poplin and Kavita Vemuri

4.1 Introduction

Striving to reach consensus about the use of resources is crucial in spatial planning. Civic engagement and participatory planning support activities of negotiation and consensus building. Negotiation, as considered in this work, is a process of communication in which parties exchange their messages, opinions, or statements in order to influence the other party (Fisher 1991). In simple terms, negotiation is a discussion between two or more disputants who are trying to work out a solution to their problem. Many situations in urban and regional planning require negotiations and consensus building. Some examples may include questions like where to locate a new road; how to design the newly created park; and what is the best location for a new shopping mall. A negotiation can be interpersonal where several individuals negotiate, or inter-group in which groups negotiate among themselves. It can include different stakeholders: the residents of the planned area, various government departments, real-estate developers, industry, and non-governmental organizations (NGO's). Reaching a consensus among different stakeholders is a challenging task which often needed to involve compromises among all involved parties. These negotiations take place because the stakeholders and individuals wish to create something new or resolve a problem or dispute. The problem usually arises when there are conflicting interests involved on how to use natural resources, land,

A. Poplin (✉)
Iowa State University, Ames, IA, USA
e-mail: apoplin@iastate.edu

K. Vemuri
International Institute of Information Technology, Hyderabad, India
e-mail: kvemuri@iiit.ac.in

© Springer International Publishing Switzerland 2018   63
O. Ahlqvist, C. Schlieder (eds.), *Geogames and Geoplay*, Advances in
Geographic Information Science, https://doi.org/10.1007/978-3-319-22774-0_4

buildings and/or how to revitalize and further develop cities and landscapes. One of the big challenges faced by planners that facilitate participatory planning and civic engagement represents the process of consensus building in which the parties can present their conflicting points of view with the goal of arriving at an agreement.

This chapter explores the possibility to use an online game-based approach for negotiations and consensus building in urban planning. In general, the geo-location could be anywhere in the world, and the game implementation might result in a spatial game, sometimes referred to as a geogame (Schlieder et al. 2006; Ahlqvist 2011; Ahlqvist et al. 2012; Poplin 2012b, 2014) or location-based game (Schlieder et al. 2006). In our case study we selected one of the largest low-income areas in Mumbai (India), which is called Dharavi. The area is inhabited, at a conservative estimate, by around 3,000,000 residents. Most of them are employed in the service sector or run their own small enterprises. Even though this is considered a low-income community, the small-scale enterprises and the skills of the people play a crucial role in the economy of the whole city. Over the last three decades, several re-development plans for the area have been proposed by the Dharavi Redevelopment Agency (DRA n.a.), real estate developers, and non-governmental organizations such as the National Slum Dwellers Organization (NSDO n.a.). Many of the plans faced resistance from the local residents who are the primary stakeholders. The governmental and city organizations did not attempt to organize a process of consensus building with local residents (Arputham and Patel 2010), but rather moved forward to with preparations to present the new master plan. This caused additional problems because some of the stakeholders were not willing to accept the suggested re-development plan.

The development of Dharavi is an interesting case study for our novel approach of using an online game-based application for consensus building; the process of exchange and negotiation within the game brings different stakeholders together to listen to different points of view. The main goal of the game **YouPlaceIt!** presented in this chapter is to enable stakeholders to communicate and resolve urban planning issues. The challenges arise when there are different and conflicting ideas about and how to revitalize and further develop cities and landscapes and how to use natural resources, land, and buildings. This game assumes that the stakeholders wish to re-develop the area, create something new, and resolve problems or disputes, and thus parties wish to negotiate. One of the main tasks of the consensus building process is to enable information exchange, communication, and the ability to express views without the fear of backlash from the community or from the powerful parties involved in the negotiation process. In a broader sense, we would like to contribute solutions which could contribute to a more sustainable way of living and co-creation of cities in which everybody feels heard and accepted.

Drawing from existing literature on negotiation and serious games for urban planning, our goal is to assess whether it is possible to develop a game that can bring different stakeholders together who are facing urban planning challenges. Section 4.2 reviews previous work related to collaborative planning and consensus building, negotiation models, communication in physical vs. digital environment, Public Participation Geographic Information Systems (PPGIS), and digital serious spatial games for urban planners. Section 4.3 introduces the complexity of urban planning process in the selected study case of a slum area Dharavi in Mumbai, India. It suggests game-based negotiation and consensus finding as a possible solution which could potentially bring different stakeholders together. Section 4.4 describes the implemented game **YouPlaceIt!** while Sect. 4.5 provides results of testing on a small sample of potential users of the game and suggests further research and implementation directions. Section 4.6 concludes the chapter with an overview and a discussion about the findings.

4.2 Previous Work

4.2.1 Collaborative Planning and Consensus Building

Collaborative planning is an interactive process of consensus building among different stakeholders and public engaged in planning activities. Collaborative planning requires participation and engagement at the bottom, or grass roots level, and a strong political and professional support at the top (Fischer 2006). The main shift towards collaborative planning happened in the 1990s and was based on communicative action theory (Innes 1995). The role of planners, the public, and stakeholders gradually changed; planners became facilitators and mediators, considering the different opinions and positions of the parties involved in collaborative planning processes (Brooks 2002). Additionally, Fischer (2006) describes active participants as empowered, participatory citizens that can effectively participate in shaping public programs and policies. An empowered citizenry is achieved when the citizens provide their input related to the topics discussed and the government intentionally pursues their input and provides the needed resources and knowledge for the citizens to participate and influence public decisions and planning policies. Communicative action or collaboration is the theoretical model for planning, while consensus building or capacity building are techniques within those models.

Consensus building in urban planning is a complex process with several stakeholders involved in planning, re-development, and decision-making. These groups are often driven by self-interest, following an agenda that can conflict and contradict the goals of other involved groups and individuals. Most proposed development plans by either government or private real-estate developers are prone to distrust by the local community. This is especially the case when the local community (a) does not have access to all data/information, (b) experiences a lack of transparency in information sharing, (c) are not involved in the planning and negotiations about the development plans which affect them, and/or (d) are not involved in decision-making.

In the process of consensus building, all opinions and concerns must be considered and seriously assessed by all involved parties. Such consensus building processes depend on the involvement of a diverse range of stakeholders and individual citizens. During the consensus building process, the involved parties can gather a variety of information about planning activities and alternative solutions and exchange their opinions about the discussed topics. The expressed opinions and experience can be used to create plans, evaluate the effectiveness of systems and projects, and adapt the processes to meet ever-evolving goals (Innes and Booher 2002). The consensus building process can be organized on a virtual platform or in a face-to-face environment. The facilitator/planner may design the process and mediate among the parties. In this way, the involved parties can explore planning alternatives, learn about them, and finally arrive at a solution which they could all agree upon (Innes 1996).

Collaborative planning can facilitate creation of sustainable communities that can effectively deal with the complex issues facing cities today (Innes and Booher 2002; Roseland 2005). This chapter will explore whether a consensus building process can be facilitated with the help of a computer-based application, and the possible implementations that could support effective information exchange and online negotiations.

4.2.2   Negotiation Models

Negotiation is a process of communication and some researchers distinguish between linguistic and non-linguistic approaches. The social-psychological perspective considers negotiations as a way to satisfy the needs of the parties involved in the negotiation process. Negotiation is also studied as a strategic behaviour (Donohue et al. 1984; Putnam and Jones 1982; Fant 1992). Sokolova and Lapalme 2012 distinguish between the means of interaction (face-to-face, email, chat), communication modes (synchronous or asynchronous), and interaction modes (one-to-one, one-to-many). Negotiation models exemplify human behavior like decision-making processes of the parties involved in the negotiation and consensus building, conflict resolution and cooperation. Which negotiation models can be applied in a serious game and how can they best be implemented? Some possible behavior models that can be applied in a game include the Nash equilibrium, Quantal response equilibrium QRE (McKelvey and Palfrey 1995), cognitive hierarchy (Camerer et al. 2004), or the level-k (Costa-Gomes et al. 2001; Nagel 1995).

The level-k model may enable to model strategic thinking and the possibility to adopt an optimal response of a player to the beliefs about other players and their activities. This model attempts to depict the closest to real-life conditions and situations. It can capture the essence of each role the player has in the game. One of the variants or conditions of the Nash equilibrium is the zero-sum game, which translates to one party wining at the cost of the other, a proposition that could have disastrous implications in many urban development plans. The level-k model combined with a game play that strives or forces people to reach mutually beneficial or agreeable consensus can be achieved by principled or integrative negotiation models, as implemented in the case study discussed in this chapter.

4.2.3   Communication in Physical vs. Virtual Space

While consensus finding in urban planning occurs mainly in the physical space through face-to-face discussion among various stakeholders trying to reach a solution; urban planning games can constitute an interesting alternative in a virtual space. These games can allow for e-negotiations through multiple channels; players can take on different roles, and varied solutions can be simulated and analyzed through a visualization of the urban space under consideration. Firth makes a distinction between activity and encounter in negotiations. He distinguishes between a *negotiation encounter*, which is physically defined as the location where the conflicting parties convene, and a *negotiating activity*, which refers to the communicative interaction of the parties involved and their aim to reach mutual alignment. The two aspects are not regarded as the same thing and furthermore are not interdependent.

Sokolova and Lapalme (2012) show that the use of electronic means changes the way people communicate during negotiations. In face-to-face negotiations, information can be gained through a non-verbal body language such as gestures and movements as well as language characteristics such as tone of voice and pauses. Language also plays an important role in text-based electronic negotiations, offering insights in the negotiation process (i.e. conditions of bargaining, introduction and closure) as well as in the social aspects. An analysis of the informativeness of messages exchanged by negotiators based on linguistic signals (i.e. the presence or absence of degree, scalar and comparative word categories) shows correlations with negotiation success or failure (Sokolova and Lapalme 2012). The use of text-based forms of communications such as online messages boards, email, instant messages and text messages, and their impact in the communication process, have been extensively analysed in the literature (Naquin et al. 2010); Clark and Brennan 1991). Considerably less attention has been devoted to the embedding of these media within (online) games (i.e. in-game chat). Instead, they have been addressed mainly in the educational context, especially with respect to language learning (Kardan 2006).

4.2.4   Public Participation Geographic Information Systems (PPGIS)

Several recent developments aim to offer alternative online participatory options. Public Participation Geographic Information Systems (PPGIS) utilize a geographic information system (GIS) as the base technology that stores spatial data and enables spatial analysis and spatial queries. The user interface is based on interactive maps displaying the area which is the main subject of participatory process. Participatory functionalities are added to the main GIS user interface; they aim to enable the stakeholders to express their opinions, participate in public debates and contribute the discussion about currently relevant issues in their neighborhoods, cities or states.

PPGIS was conceptually developed in the 90s, followed by example implementations in the mid-90s. The term PPGIS was coined in 1996 at the National Center for Geographic Information and Analysis (NCGIA) Workshop, Orono, Maine, July 10–13, 1996 (Schroeder 1996). The concepts of PPGIS and its possible implementations has been intensely discussed by many researchers (Pickles 1995; Schroeder 1996; Rinner 1999; Talen 1999; Kingston et al. 2000; Al-Kodmany 2001; Basedow and Pundt 2001; Carver 2001; Jankowski and Nyerges 2001; Craig et al. 2002; Schlossberg and Shuford 2005); Georgiadou and Stoter 2010). In the mid-90s, PPGIS promised a novel way of citizens' involvement into decision making enabled by online, interactive maps and included participatory functions. These participatory functions allowed posting online comments related to spatial objects, which lead to the concept of argumentation maps introduced by Rinner (1999, 2001, 2005, 2006). They enabled online chats, comments, discussions, sketches, and exchanges of opinions related to the issues in question. The citizens were able to send their annotated maps to the planning authorities (Steinmann et al. 2004) or leave a comment directly on the online maps (Rinner 1999; Al-Kodmany 2001; Rinner 2005, 2006; Poplin 2012a, 2015). These comments were then stored in a GIS database. A GIS enabled to display them according to the criteria needed for spatial analysis. Figure 4.1 shows an example of a PPGIS applications developed to discuss ways inhabitants of Wilhelmsburg, in the city of Hamburg, use their canals (Poplin 2012a). The three pencils on the right hand side of the map enable them to draw paths they use for walking, jogging, walking their dog, biking, and other activities.

The idea of PPGIS resulted in a variety of implemented applications aiming to enhance participatory processes with novel digital visualization possibilities. An

Fig. 4.1 PPGIS for Wilhelmsburg, Hamburg, Germany (Poplin 2012a)

intense debate in the scientific research community lasted about 15 years (90s to mid-2000s). In 2006, Sieber published her paper providing a thorough overview of the research contributions looking back at more than fifteen years of research effort. The positive aspect of PPGIS development was summarized by Shuford (2005) as follows; "PPGIS represents a broad notion that the spatial visualization and analysis capacities inherent in GIS present a unique opportunity for enhanced citizen involvement in public policy and planning issues." Other advantages are related to the data. Data presented and collected online is georeferenced, stored in a digital format and easier to process and analyze than data collected at the traditional public participatory meetings (Kingston et al. 2000). It can be shared easily online (Thompson 2000) and is accessible by many at the same time from anywhere with an internet connection.

The critique of PPGIS focuses mostly on the usability of these complex and technically advanced systems (Haklay and Tobón 2003). Steinmann et al. (2004) and Poplin (2015) report on the complexity of the applications and the problems users might have while using them. The idea and the conceptual model did not result in many practically successful PPGIS applications. The problem of complexity of PPGIS implementation and the lack of their user-friendliness lead to novel ideas such as serious digital geogames, often referred to as spatial or location-based games. These terms will be used interchangeably in this chapter. Serious spatial games aim at overcoming the issues of PPGIS, especially due to the complexity of the user interface introduced to the users, while still using GIS as the technology that enables visualization of interactive environments.

4.2.5 Digital Serious Games for Urban Planning

Digital serious games may enable online experimentation, online exploration of the urban planning situation, visual and interactive representation of the key issues that are being discussed in the planning process, and alternative ways of collecting data and involving citizens into planning for the future of their cities. With their focus on serious issues, they can be categorized as serious games, games designed for "more than just fun", "entertaining games with non-entertainment goals", or games for change. They may aim to facilitate learning (Gee 2004, 2005; Lemke 1998) and problem solving (Abt 1970). Learning in such games can be facilitated with the help of experimentation (Lemke 1998); several alternatives can be tested in the game environment without serious consequences, which would happen in the case of a non-game, real-world experience. The players may learn, study the information given to them, in the order that suits them and at the speed and pace that can be optimal for their own personal experience. Digital serious games can come in "any form of interactive computer-based game software for one or multiple players to be used on any platform that has been developed with the intention to be more than entertainment" (Ritterfeld et al. 2009, p. 6).

Recently this area of research gained more attention due to a possible combination of GIS and games. Several digital serious games for urban planning have been developed. Gordon and Manosevitch (2010) describe a pilot project "Hub2", which took place in Boston, Massachusetts from June to August 2008. Devisch (2011) focuses on Second life and its possible use for urban planning. Poplin (2012b, 2014) introduces the NextCity and B3-Design your Marketplace! games, which aim to facilitate public participation in urban planning. None of these games focused on negotiations and consensus building. They mostly aim to support civic engagement and explore the possibilities for the implementation of game-based, playful civic engagement in urban planning. We are interested in how consensus building can be implemented in a game-based online system. Can serious spatial games for urban planning enable and facilitate consensus building in civic engagement situations? How can they best support consensus building processes? The next section introduces the study case Dharavi and the main requirements for an online game-based negotiation and consensus building.

4.3 Negotiation for Building Consensus Among Stakeholders: Case Study of Dharavi

4.3.1 The Selected Study Site: Dharavi in Mumbai, India

Our specific focus is on the situation of Dharavi, one of the biggest slum areas in India. It was declared by the Slum Rehabilitation Agency (SRA) as one of the most difficult areas to revitalize due to many conflicting interests about how this site should be further developed. Dharavi is embedded into the city of Mumbai, and is located between the Sion Hospital on the south, surrounded by three major roads on the north-east and east (Fig. 4.2). It is accessible from both Western and Central railways. With the advantageous location neighbouring the business district in Mumbai, the land of Dahravi is of premium interest among builders.

Dharavi offers home to more than three million residents who can be described as self-sufficient people. The inhabitants of Dharavi are also known for their entrepreneurial spirit. They create job opportunities on their own, which range from small shops to the citizens involved in recycling activities (Fig. 4.3a, b). Dharavi is home to more than 15,000 small-scale home businesses. The recycling units of Dharavi generate revenue by turning around the discarded waste of not only Mumbai's 21 million citizens, but from all around the country and abroad. However, the government does not recognize these industries as it would have to start giving subsidies (Chandan 2014).

4.3.2 The Issues of Civic Engagement in Dharavi

The redevelopment project of Dharavi was first announced in 2004, followed by the presentation of the new master plan. The main goals of the redevelopment plan can be summarized as follows: formulate sustainable development master plan,

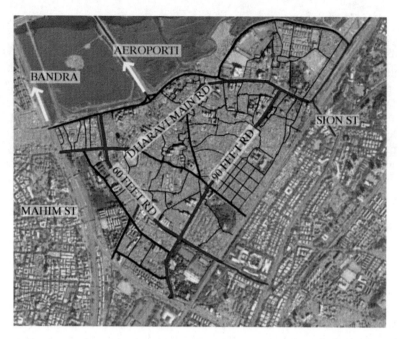

Fig. 4.2 Google satellite image of the study area of Dharavi in Mumbai, India with proposed new road expansion plans

Fig. 4.3 (**a** and **b**) Small entrepreneurs of Dharavi (Chandan 2014)

rehabilitate all the slum families/businesses, retain all eligible existing rehabilitated families/businesses, rehabilitate non-polluting industries, and integrate slum dwellers with the main stream residents (Mehta 2010). In 2007, global tenders were invited to submit an expression of interest to be part of the redevelopment activities. Slum Rehabilitation Agency (SRA), together with consulting teams, offered high incentives to the citizens and other stakeholders to comply with the solutions proposed in the master plan. Over 200 PowerPoint presentations that promoted the presented master plan were shown across the area. They were organized for the local inhabitants, local leaders, political leaders, and leading entrepreneurs. About 70,000 pamphlets and 500 posters were distributed in Dharavi to inform the stakeholders about the planned activities. The organizing institutions/organizations included Maharashtra Housing and Development Authority (MHADA), Slum Rehabilitation Agency (SRA)—the proposed rehabilitation and implementation agency at that time, the Local Corporators implementations agency at that time, the Municipal Corporation of Greater Mumbai (MCGM), Mumbai Metropolitan Regional Development Authority, Assistant Director of Town Planning, the Chairman of the co-operative societies of Dharavi, the NGO's operating in Dharavi, Department of Housing/Government of Maharashtra, Department of Urban Development, Department of Finance, the Chief Secretary/GoM, and the representatives of the Planning Commission. The process considered by government agencies has been to make unilateral decisions. Politically and socially this lead to a sense of distrust by the community and individual owners, exclusion of many stakeholders, and many disagreements about how the space should be used, re-vitalized, and further developed.

Following the global real-estate crisis in 2008, five executing companies exited the project in 2009, citing lack of clarity and delays in the implementation. Some of the key-stakeholders and bidding companies withdrew from the project as a result of dissatisfaction, lack of transparency, lack of information sharing, and support of the plan by the local inhabitants (Mehta 2010). Of 14 bidders, only seven submitted the Memorandum of Understanding they signed with their foreign partners. In the same year, a survey report claimed that 63% of Dharavi residents are ineligible for houses defined within the project. In spite of all the talks, presentations, flyers, and posters, only 15% of Dharavi was restructured and redeveloped in the period of 15 years. The local developers kept building poor quality housing in this area with negligible benefits to the local citizens, entrepreneurs, landowners, and government. Even as the state government finally begun to work on the Dharavi redevelopment project, residents of the area lacked clarity about the status of the project; they did not feel informed about the redevelopment plan.

4.3.3　Novel Approach for Civic Engagement in Dharavi

The entrepreneurial spirit of Dharavi and the active inhabitants of this redevelopment area may be attracted by fresh approaches to citizens' engagement and public participation in urban planning. Novel digital technologies can enable creation of online platforms for discussions, debates, and a method of consensus building by inclusion of all stakeholders. In our approach we suggest a development of a serious, online, digital spatial game which may focus on sharing the information about the redevelopment of Dharavi and enable the stakeholders to communicate and exchange their positions on the important issues and decisions to be made in the redevelopment process. The users of a game may be the inhabitants of Dharavi, and organizations and agencies such as Maharashtra Housing and Development Authority (MHADA), Slum Rehabilitation Agency (SRA), the Local Corporators implementations agency, the Municipal Corporation of Greater Mumbai (MCGM), Mumbai Metropolitan Regional Development Authority, the Assistant Director of Town Planning, the Chairman of the co-operative societies of Dharavi, the NGOs operating in Dharavi, the Department of Housing/Government of Maharashtra, the Department of Urban Development, the Department of Finance, the Chief Secretary/GoM, and/or the representatives of the Planning Commission. The digital technology may enable information sharing and dialogue among the involved stakeholders.

Our suggested approach includes a development of a prototype for an online game that may enable negotiations related to the planned activities in Dharavi or any other selected place/neighborhood. The communication may be based on the use of language and creation of online, digital environments that may enable consensus building. Can we create an online serious geogame that will enable negotiations? Can such a game bring people of Dharavi and involved stakeholders together? Which negotiation algorithms can be used in such situations and how can they be implemented within a game? What are possible limitations of this approach?

4.3.4   The Main Requirements for the Negotiation and Consensus Building Game for Dharavi

The main requirement for a negotiation game for Dharavi focuses on bringing all parties interested in redevelopment of Dharavi together. The game aims to serve as a platform for an exchange of information, an exchange of opinions, and often contradicting ideas about the use of space in Dharavi. The main goals of the game are to enable the users to communicate, exchange their opinions about the proposed projects, negotiate about the proposed projects, and view the progress of discussions and the implementation of the redevelopment plan.

The requirements for the game design process may include the following guidelines:

- The game represents a specific, concrete area that exists in the real world: Dharavi can serve as a very specific example where civic engagement and public participation are needed in order to discuss and view the redevelopment plans.
- Relation to the real-world issues/area represent the "serious" part of the game: The visualization of Dharavi should be realistic in order to enable focused discussions about very specific projects and redevelopment plans.
- Fun to play: The game aims to attract many different stakeholders. The fun aspects of the game can be implemented in a variety of different ways. The game can offer fun characters, elements of competition, a budget to be invested

in projects, or just the possibility to build a community focused on topics users care about.

- Game is online and uses geospatial technology: The online accessibility of the game is attractive due to the availability of smart phones owned by many inhabitants of Dharavi. It can also be attractive for many small entrepreneurs due to the flexibility of the time when the game can be used/played 24/7 with internet connection.
- Option to negotiate integrated in the game: The negotiation mechanisms can provide an opportunity for an exchange of opinions related to the very specific projects suggested by the stakeholders.
- Negotiation language integrated in the design and implementation of the game: The language enables the users to express their opinions in a more natural way, in the way that feels appropriate in case of conflicting positions. The language expression may be combined with more specific, quantitative measures such as the amount of money that can be invested in a project.
- Supports collaboration among the players: The collaboration can be achieved in a form of a dialogue in which conflicting parties exchange their opinions. This dialog can be supported by additional facts about the planned project such as the budget available, the land use regulations, and others contributed by the experts in a collaborative discussion with the inhabitants of Dharavi.
- Disagreements among parties can be resolved by negotiations: The ability to exchange opinions and facts can provide a platform for a serious, fact-oriented discussion in which everybody involved feels appreciated and heard.

A negotiation game aims to support building and co-creating sustainable cities which may enable everybody involved to express his/her wishes, explore about the possibilities for changes, and learn about the consequences of implementations while co-creating living environment together.

4.4 Serious Digital GeoGame for Negotiations and Consensus Building: YouPlaceIt!

Our main goal is to explore possibilities for the design and prototype implementation of an online game that can enable online negotiations and consensus building in urban planning. We implemented a test version of an online gametitled **YouPlaceIt!**. It was implemented by the team at the International Institute of Information Technology. The game is one of a kind; we are not aware of any similar product on the market. It falls into a similar genre presented in **B3-Design Your Marketplace!** game (Poplin 2014), which enables the players to design their own urban space by using different objects such as benches, trees, lights, fountains, etc. By placing these objects in 2D or 3D game space, they can develop their own spaces, just the way they would like them to be in the real-world.

4.4.1   Goal of the Game

The main goal of the game is to enable negotiations about the future use of space and provide the needed transparency for the decision-making process in urban planning. It aims to support the exchange of opinions related to a specific building project expressed by different stakeholders. The stakeholders involved are individuals living in this particular area, builders, contractors, urban planners, and other public officials. The game aims to enable an exchange of player's/user's views, opinions, suggestions for changes, and focuses on the negotiation algorithms. Additionally, the goal of the implementation was to create a digital online prototype of the game; a digital platform that allows for testing of the user interface, the gameplay, different scenarios for changes in the environment, and to validate the responses submitted by the players.

4.4.2   Premise

The premise of the **YouPlaceIt!** game focuses around a road construction process. Imagine different stakeholders having their self-interest driven utilization ideas on how a certain space/piece of land in a city can be used. They have their own interests where the new road should be constructed, how much money should be invested into this particular project, and who will the involved parties be. These ideas can be in conflict or contradiction with the ideas of other stakeholders involved into the gameplay. Stakeholders can invest a selected amount of money into a road construction project; the involved players can respond to that with their evaluations of the worth for this particular investment. Some other players might completely oppose to the project, find it too expensive, or disagree with the proposed location of the road construction project. The players can negotiate the price for buying/selling properties that lay along the planned road. The main idea is to enable an exchange among the players, facilitate negotiations about the monetary value of the building project, and to reach a consensus, a win-win outcome for all involved in the discussion. The gameplay is limited to a road construction project, but can be expanded for other kinds of projects.

4.4.3   Game Elements

Space/Environment. The space represented in the game can be any space one wishes to visualize as raster data; in this case satellite images. In order to illustrate the complexity of the planning process, we present the aforementioned case study of Dharavi, an area of about 230 hectares (approximately 557 acres) with a contentious urban planning history. Figure 4.2 shows the area on a satellite image zoomed in from Google maps in a 2D representation.

Players. Presently a two-player game is implemented; in the future we intend to implement a multi-player game. One of the players can take on the role of a local community representative who can tag the sites as residential, religious structures, playgrounds, hospitals, or schools. With tagging, the player can identify the main buildings in the area. The player can also enter the expected sale price per sq meter and the margin for negotiation around which the discussions and negotiations evolve. The second player can take on the role of a road developer, which in this case could be a government agency or a contractor entrusted with the construction. We envision the game to be played by the actual inhabitants and the government agencies entrusted with the road development. These players possess the local knowledge required to be able to play the game.

Objects. The current version of the game focuses on road objects and the possibility to tag the objects, to build new roads, and renovate/change the existing roads. The objects representing the properties are also included in the game. These objects can be tagged by color-coded pins classified as either government (hospitals, schools, or parks), non-governmental (places of worship), or private (houses or shops). The color codes for each categories of property in this version are as follows: NGO property use (pink tag), government property (green tag), and private property (blue tag). The road object can be visualized on the satellite image. In addition, there are other objects that represent various categories of existing urban structures which cannot be moved, demolished, or tagged by the player playing the local community representative. An additional object includes a drop-box for selecting the property for negotiation, input the price, and a chat-box for informal discussions.

Resources. Current resources in the game are money and space. Money is limited by the budget available to the players in their negotiation process. The game starts with this budget suggested by a player. The suggested budget can be allocated to the cost of building a newly suggested road. Space is limited by the ability to build the road in certain areas, but not in others.

4.4.4   Gameplay

The game starts with the road developer indicating the available budget for the road construction project. Figure 4.4 shows the user interface for inserting the initial suggested budget. The initial budget is not limited by any means; the player can insert any number he/she finds reasonable for the planned project. The player taking the role of the road planning agency can suggest a new road to be built by selecting the path of the road on the satellite image. The suggested road construction can be changed or deleted later in the process of finding the optimal route for the new road.

Figure 4.5 presents the pop-up box that enables to insert the shape of the buffer zone, name of the property and type (hospital, temple, or school), radial distance (only if the shape selected is a circle), the selling rate per square meter, and the negotiation margin. The negotiation margin can be inserted in percentage. For each

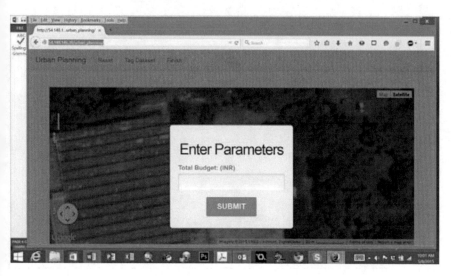

Fig. 4.4 User interface for inserting the allocated budget for constructing the new road

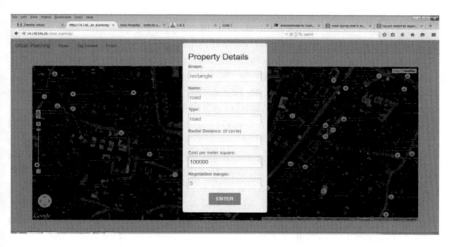

Fig. 4.5 The pop-up text box that enables basic input

of the property types, a perimeter/radial distance is specified as a buffer zone, which ensures that a road is not constructed too close to the buildings such as hospitals, schools, or religious structures. A per square meter price of the land for acquisition is also specified plus the negotiation margin in percentage.

The road planner can indicate the tagged area on which he/she would like to build a road (Fig. 4.6). The player can than draw a path connecting two points indicating the proposed road position by selecting the 'draw path' option on the menu bar. Considering that the area is densely populated, the road can be constructed only

Fig. 4.6 Places identified as different types of properties

after the property has been acquired. The player taking on the role of a the local community representative can label buildings and open spaces like parks and places of worship on the satellite image. Each of the properties can be owned/represented by a non-governmental agency (temples, hospitals, or schools), a private owner (residential, hospitals, or schools) or a government owned property (schools, hospitals, power stations, water distribution control centers, etc.). The color code for each categories of property are: NGO property use (pink tag), government property (green tag), and private property (blue tag). Figure 4.6 shows the places identified as NGO property, government property, and private property. The patch in red is the buffer zone marked by the player around the property.

Figure 4.7 shows a set of tags selected over the Dharavi area where three large government properties are tagged. At the end of the tagging and selection process, the major properties in the indicated area are tagged, and a table with the inputs on price is generated and a buffer-zone area is created for further calculations.

To mark the buffer zone around the property, a rectangle or circular region of interest can be drawn. A rectangular shape with latitude/longitude coordinates or a radius in meters for a circular shape can be selected. A drawing tool is provided for a player to trace the path of the planned road. An algorithm to calculate the cost for land acquisition based on the property prices can be run by selecting the 'process path' toolbar button. The total cost of all acquisitions can be viewed by selecting the 'construction cost' button. A pop-up displays the acquisition cost, the initial budget outlay and the number of negotiations remaining (Fig. 4.8).

The green tags on Fig. 4.8 indicate two government properties close to or on the pathway and the red patch represents the buffer zone. In order to be able to acquire these properties, the budget required for the transaction is displayed by selecting the

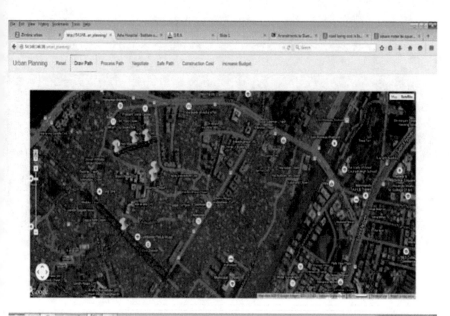

Fig. 4.7 Six governmental properties are tagged

Fig. 4.8 The proposed road path shown as a black line with red/yellow intermediary markers

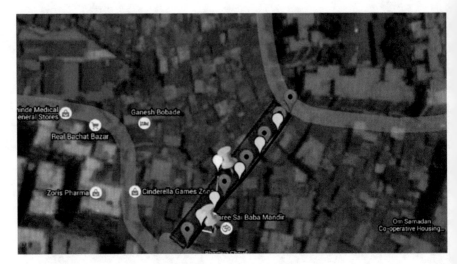

Fig. 4.9 The proposed alternate road segments are displayed as 'white' markers

'construction cost' button. The 'construction cost' button appears as a pop-up text box on the top right corner. The road planner is given two options. The first option: No negotiation—that is, consider the optimal path generated by the system. This can be done by selecting the 'safe path' button. An equation considers the buffer zone areas input by the other player and suggests localized diversions, that is, small changes in measured in meters or feet from the original path to circumvent the costs or social issues for removing an existing structure like places of worship or a public utility building/space. The construction budget can also be increased by clicking on the 'increase budget' button. The second option: Price negotiations can be initiated if the initial pathway is most optimal and localized diversions would lead to future issues, which could be other property owners who have objections or conflicts within government agencies like the waterworks department or utilities department.

Figure 4.9 shows the alternative paths and displays them as 'white' markers. The broad 'red' strip is the buffer zone around small residential houses, the 'black' line is the path for the road drawn by the road planner, and the 'green' tags are labeled properties with their corresponding buffers. Based on buffer parameters provided, the system calculated small detours the path should take to avoid a conflict in land use or price escalation.

4.4.5 Negotiation and Communication

YouPlaceIt! online digital platform focuses on enabling a dialogue among playing parties involved in negotiations and consensus building processes. The game is designed around the concept of *principled negotiation*, in which the players are

Fig. 4.10 Negotiation box on the *top right* and the chat box *below* for informal messages

encouraged to close the deal with mutually beneficial financial positions. The negotiation is based on fixed margins suggested by the negotiation parties. The negotiation model is based on the premise that all stakeholders desire a change in present conditions of living and are willing to reach a consensus, which is possible if the players understand the beliefs and optimal decisions the other player will make. In order to enable a free dialogue and text exchange among negotiating parties, we implemented a chat box. The input given by the players can further enable a more complex analysis of the negotiation and consensus building processes.

Figure 4.10 shows the implemented negotiation box. The negotiation box on the right side of the user interface allows the player to input a price he/she is willing to pay for the property tagged in 'green'. The price negotiation can start by clicking on the 'negotiate' option on the menu bar, which opens a panel on the right-side of the screen. A drop-down box lists the number of properties requiring price negotiations and the road player can select them one at a time and enter a price that he/she is willing to pay. A price as per the market rate and a profit expectation of the community player is calculated from the per square meter cost input which also includes the buffer zone. For example, if the per square meter cost was 100,000 and the buffer area was 50 m^2 than the starting price of this property will be "area × cost-per-sq meter" in this case 5,000,000. If the road planner inputs a price lower than the cost of acquisition of the above property, the new price is calculated from the negotiation percentage provided by the property owner. For example, if the margin was 10%, than the new cost will be calculated as follows: actual_cost × profit margin/100. This new value of the property will be displayed on the right-side window panel. This negotiation process can continue until the road planner agrees on the sale price proposed.

An informal chatbox is included for the player who tags the properties and for the road planner. It enables them to exchange informal messages in the form of a free text. An analysis of these chats can provide interesting data that can enable a better understanding of the language-based (qualitative) negotiations in addition to the price-based (quantitative) negotiation.

4.5   Testing YouPlaceIt! Game and Reflections About the Implementation

The game was developed according to the user-centered design approach placing the user in the center of the development. During the initial prototyping phase it was tested in Utrecht, Hyderabad, and Ames, with internal researchers working on the project. Several revisions were made in the process of **YouPlaceIt!** game development. The second phase of testing was executed in Hyderabad with a limited number of players.

4.5.1   User Experience Tested with the Selected Players

A focused testing was executed with two architects and a civil engineering consultant. The number of test players was small, but gave enough feedback for the revision of the game to achieve an improved version of the game prototype. The majority was accomplished in an open interview with the players summarized in this section.

The two selected architect users/players included a female with 5 years of experience and a male with 10 years of experience in their profession. The civil engineering consultant was a male with 20 years of academic experience and experience in industry consultancy, mostly in construction. All three of them are regularly involved in urban planning activities and processes. They were given access to the online platform **YouPlaceIt!** and the possibility to experiment with it. We prepared ten questions with yes/no/maybe being possible answers to these questions. The results of this testing are not representative but can indicate certain trends, open up questions for a discussion, and give us directions for our future explorations and research. The questions with their corresponding answers are summarized in Table 4.1.

From the small set of answers followed up with open discussions about **YouPlaceIt!** communication and negotiation game we can infer that an online tool that allows access to the information, includes options for exploration, and an exchange of opinions is certainly needed. Another positive aspect was the perception of transparency and the feeling of fairness that can be conveyed via such applications. Reflecting upon the use of informal chat-box negotiations, one of the participants noted that deciphering language, especially in cultures like India where every regional language is complex and differs from other regions, might confuse

Table 4.1 Questions with corresponding answers

	Question	Architect 1	Architect 2	Civil engineer consultant
1	Are you able to understand the aim of the application?	Yes	Yes	Maybe
2	Have you played games which have a sole aim of teaching/instructing on a specific concept? (example: first-aid methods, fire-evacuation, culture training etc.,)	Yes	No	No
3	Can public participation in urban planning contribute to a more inclusive development of the city/nation?	Yes	Yes	Yes
4	Do you consider online public medium like simulation games as serious/ valuable online instruments for the collection of citizen's opinions, desires?	Yes	Yes	Maybe
5	Do you prefer negotiation on real-estate matters by electronic medium? 5a. Do you prefer face-to-face negotiation?	Maybe	Yes	No 5a. Do you prefer face-to-face negotiation? *Social conditions can be better gauged in face-to-face. Inclusion of video/audio chats.*
6	Can the process of a property owner (NGO's, private, government) tagging their property and indicating a sale price lead to transparency in real-estate deals?	Yes	Yes	Maybe
7	Is the process of each property owner (NGO's, private, government) tagging their property and indicating a sale price lead to property owners getting fair deals for the government?	Yes	Yes	Maybe
8	From a cultural perspective, do you think a medium such as the one proposed would be considered seriously by government agencies and/ or citizens?	Yes	Maybe	Maybe
9	Do you think inclusion of informal discussion by chatbox between stakeholders can increase the engagement?	Yes	Yes	No. *Consider the language used which can differ from participant to participant, sometimes the chat cannot be moderated, the informal chats might confuse some stakeholders and lead to legal litigations.*

(continued)

Table 4.1 (continued)

	Question	Architect 1	Architect 2	Civil engineer consultant
10	List the major changes required to make this an acceptable tool? (These should not be suggestions on user-interface or navigation as in the current prototype).			*a. Embed already labeled and tagged property information from government agencies and citizens.* *b. Include other exercises which facilitates better coordination between intra-governmental agencies—road works, telecommunications, water works, etc.* *c. Include a moderator of the discussion/ negotiation*

the players and lead to intransparent negotiation processes where several possible languages could be used in one negotiation topic/process. The role of a moderator was discussed; a moderator of the discussion/negotiation could additionally clarify questions and lead the discussion in cases which might not be clear to some of the participants/players. In the later stages of implementation, a discussion and focused attention will be devoted to this suggestion.

The replies from the participants to locational queries focused on a particular project/topic of interest were evaluated as positive; they can initiate discussions among involved stakeholders and, stored into a digital database, they would be relatively easy to analyse. There was a general consensus among the three testers that validation of the inputs accomplished by the property owners is required which should be provided by a neutral body like a regulator or environmental protection agency. A game is a playful approach that can reach many stakeholders. Due to the serious matter discussed, a validation procedure needs to be included in the process to assure the accuracy of the inserted data. Several additional, more playful elements may be added to the current concept of the game and may include fun characters, random events, levels, and/or scoring mechanisms.

In the future we intend to test the game with different user's groups in an international setting and explore how the game design, game mechanics and user interface influence the gameplay. Testing may lead to further suggestions for the improvements of the game implementation. The implementation as presented here is the first version of the operational online, multi-player geogame which aims to enable negotiations and facilitate consensus building in urban planning projects. We did not come across any similar online application and consider this approach very innovative and one of a kind at the moment.

4.5.2 Discussion About YouPlaceIt! and Further Research and Implementation Directions

The communication and negotiation in the **YouPlaceIt!** game is based on the exchange of the price tags and the use of language as a communication tool. A context that is perhaps closer to the one sketched in our game is the one presented in

Rusaw, in which the linguistic practices of a community involved in the online game **World of Warcraft (WoW)** are analysed with a special focus on the discourse strategies. The main method of communication in **WoW** is a real-time text chat that is being employed to negotiate the in-game social world. Each player has an access to multiple text chat channels (from temporary to relatively permanent) which range in size from two players to several ones, therefore each player is a member of multiple simultaneous conversations. Such multiple communication channels could be potentially implemented in the **YouPlaceIt!** game in the next version of the prototype.

Further research of interest is related to the intercultural use of English as a foreign language in communication, often used in negotiations in some countries (examples include India, and also Europe). In India, for example, English is often used as the official language for public communications, in spite of not being a native tongue for many of possible involved negotiators/players. In Zeng and Takatsuka 2009), user chat logs within a virtual world are analysed and the concept of Negotiation for Action is introduced to explain how interaction between native and non-native English speakers allow for language acquisition. In particular, communication tools such as chat and email as well as 3D avatar interaction in a virtual space are employed by intercultural dyads which enable the players to negotiate and solve content related problems in English. Thus interactions involve negotiations related to the meanings, which are exchanged in a collaborative setting. The main goal is concentrated around language learning. This creates several interesting research issues related to the use of language in negotiations, the meaning of different words, and language-based exchange, which can be useful for further development of the **YouPlaceIt!** game.

The role of power is another interesting research area. Does a game like **YouPlaceIt!** support and empower one group over the other? How is negotiation modeled and implemented? Much of the discourse that takes place among the players in **YouPlaceIt!** focuses on reaching solidarity/consensus within the group involved in order to share their history, knowledge and a sense of belonging. At the same time, the solidarity of the in-group can be used to generate power over the larger number of members that are part of the out-group: power can be thus generated almost entirely by linguistic means. This is reflected also in the chat style where leaders take turns that are usually longer than normal and follow power language more likely to gain influence, expressing a more authoritative role as the information providers. Language is thus more important in this type of online game than in the real world since it is fundamental in the creation of roles and online identities of the players. Further research is needed in order to understand how is language used in **YouPlaceIt!** and whether its implementation might lead to power expressions and overpowering one player or groups of players over others.

In the future version of **YouPlaceIt!** we intend to enable multilingual chats. Multilingual text and voice chats and icons can be used as a way of communication among players in the process of reaching consensus. This set of qualitative "soft data" can be analysed to assess how the language and a dialog among the players can be employed to better understand social roles and social groups within the game. In particular, we aim at analysing whether the mechanics of the game and the identities of the "real world" have an impact in determining social roles. We plan to investigate whether there are differences and similarities in the language communication within the digital and the physical urban context setting. Language can play an important role in identifying cultural differences between social formations, groups, and individuals. This seems to be the case also for digital games in which social roles are created through language with the support provided by the dynamics of the game, while the physical identities one has in the real world play a marginal role.

Online consensus building games can be of relevance in assessing whether certain spatial configurations (physical vs. virtual) can support negotiation and consensus finding better than others so that communication and knowledge exchange can be facilitated. It is important to take into account how virtual activities influence new conceptions of physical layouts and the impact that this new vision of the physical space can have on the innovation process. More specifically, it becomes possible to analyze whether there are differences in the nature of the negotiating actions that arise in the physical space vs. the 2D or 3D virtual representation implemented in a digital online game.

4.6 Conclusions

This chapter introduces the **YouPlaceIt!** Game, which aims to support negotiations and consensus building in complex urban planning situations. The complexity arises from the variety of different stakeholders involved in the collaborative planning process, the complexity of the spatial issues related to the land use, communication with a variety of educational, income, or nationality backgrounds in the community, and sometimes unsolved/unknown property ownerships. The case study for **YouPlaceIt!** is taken from India and focuses on a very diverse slum area called Dharavi. The complexity of the situation can be additionally increased by the heterogeneous players of the game representing their own interests and visions of how this study area can be revitalized. The game introduced in this chapter is based on the idea that a game, with its playful and experimental environment, can provide a "safe space" in which people/players can experiment with different realities, interventions, and changes in the represented space. In this way they can collectively reflect upon these changes, get involved in negotiations with other players, and with this contribute to the consensus building process. The next prototype of the game may introduce additional playful elements including levels, fun characters, and internal measures of success and building the community which may enable the players to continuously keep on returning to the game.

Introducing negotiation and the use of language in negotiation is central to **YouPlaceIt!**. The game constitutes the shared space for the negotiation encounter with the 2D representation of the urban planning context on satellite images representing the context of the place to which comments refer to. Text-based chatboxes are used to enable the negotiating activities and dialogues among the involved stakeholders. They set the basis for the emergence of new identities and roles that are created through the language used in negotiations. The channels of communication (i.e. digital game) can influence the shape that the conversational process can take on. The main challenge in this research is to relate the reality of negotiations with a simulated world and modeled algorithms of negotiation. Bringing fun elements into the game and combining them with a more serious situation is an additional inspiration and a challenge at the same time.

The work presented in this chapter opened up some very interesting research challenges and topics including: the role of language in spatial games and its power in creating coalitions and power relations, the role of negotiations in spatial games and in consensus finding participatory processes, modelling negotiations in spatial games, the issues of serious aspects and how to combine them with playful and experimental elements of the game, the forms and implementations of collective reflections about the urban planning issues and how can they be facilitated by a spatial game, and characteristics of spatial games that can successfully support community planning and civic engagement processes. These questions emerged during discussions about the design and implementation of the **YouPlaceIt!** game.

In the future we plan on working on these issues and organizing our continuous testing of the latest prototypes of the **YouPlaceIt!** game. We believe that this is just the beginning of our interesting journey, which opens up several very inspiring research challenges, and contributes to creating smart, resilient cities in which everybody can get involved in the co-creation process.

Acknowledgments The authors appreciate and acknowledge Karan Damle and Jayesh Lahori, undergraduate students at the International Institute of Information technology, Hyderabad, India who developed the complete game application. This research would have not been possible without their effort. A special thanks to the local NGO, be the locals, working in Dharavi for giving insights into locals' perceptions or real-experiences when it comes to development and importantly for articulating the aspirations. The serious game development was partially funded by Department of Information Technology, Govt. of India under the National Program on Perception Engineering project Phase II, where the second author of this chapter is a principal investigator. Thank you to Brandon Klein, Iowa State University, for the language improvements of this text.

References

Abt CC (1970) Serious games. University Press of America, New York
Ahlqvist O (2011) Converging themes in cartography and computer games. Cartogr Geogr Inf Sci 38(3):278–285
Ahlqvist O, Ramanathan J, Loffing T, Kocher A (2012) Geospatial human-environment simulation through integration of massive multiplayer online games and geographic information systems. Trans GIS 16(3):331–350
Al-Kodmany K (2001) Online tools for public participation. Gov Inf Q 18:329–341
Arputham J, Patel S (2010) Recent developments in plans for Dharavi and for the airport slums in Mumbai. Environ Urban 22(2):501–504
Basedow S, Pundt H (2001) Braucht bürgerbeteiligung in der planung GIS-funktionalitäten?, CORP conference, Vienna, Austria

Brooks MP (2002) Planning theory for practitioners. Planners Press, Chicago, IL

Carver S (2001) The future of participatory approaches using geographic information: developing a research agenda for the 21st century. ESF-NSF meeting on access and participatory approaches in using geographic information, Spoleto, Italy

Clark HH, Brennan SE (1991) Grounding in Communication Excerpt: from Perspectives on Socially shared Cognition, Resnick LB, Levine JM, Teasley SD (eds) American Psychological Association, p. 127–149

Craig WJ, Harris TM, Weiner D (2002) Community participation and geographic information systems. Taylor and Francis, London

Camerer CF, Ho TH, Chong JK (2004) A cognitive hierarchy model of games. Q J Econ 119(3):861–898

Chandan V (2014) Dharavi: not a slum, but Asia's largest small-scale industry, Web. http://www.thealternative.in/society/photo-story-dharavi-not-a-slum-but-asias-largest-small-scale-industry/, 22 Jan 2014. Accessed 8 March 2016

Costa-Gomes M, Crawford VP, Broseta B (2001) Cognition and behavior in normal-form games: an experimental study. Econometrica 69(5):1193–1235

Devisch O (2011) Sollten Stadtplaner Computerspielespielen? In: Bauwelt 24.11, Hasselt. p 26–30

Donohue WA, Dies ME, Hamilton M (1984) Coding naturalistic negotiation interaction, Human Communication Research 10(3):403–425

DRA n.a. Dharavi redevelopment agency, Web. www.sra.gov.in/pgeDharaviUpcoming.aspx. Accessed 22 Feb 2016

Fant LM (1992) Analyzing negotiation talk – authentic data vs. role play in A. Grindsted A, Wagner J (eds) Communication for Specific Purposes, Tuebingen, Gunter Narr Verlag, p 164–175

Fisher R (1991) Negotiating power: getting and using influence. In: William Breslin J, Rubin JZ (eds) Negotiation theory and practice. Program on Negotiation Books, Cambridge, pp 127–140. 128

Fischer F (2006) Participatory governance as deliberative empowerment: the cultural politics of discursive space. Am Rev Public Admin 36(1):19–40

Gee JP (2004) Situated language and learning: a critique of traditional schooling. Routledge, London

Gee JP (2005) Why video games are good for your soul: pleasure and learning. Common Ground, Melbourne

Georgiadou Y, Stoter J (2010) Studying the use of geoinformation in government—a conceptual framework. Comput Environ Urban Syst 34(1):70–78

E, Manosevitch E (2010) Augmented deliberation: merging physical and virtual interaction to engage communities in urban planning. New Media Soc 13(1):75–95

Innes J (1995) Planning theory's emerging paradigm: communicative action and practice. J Plann Educ Res 14(3):183–189

Innes J (1996) Planning through consensus building: a new view of the comprehensive planning ideal. J Am Plann Assoc 62(4):460–472

Innes JE, Booher DE (2002) Collaborative planning as capacity building: changing the paradigm of governance. Paper prepared for the Association of European Schools of Planning Conference

Jankowski P, Nyerges T (2001) Geographic information systems for group decision making. Taylor and Francis, London

Kardan K (2006) Computer role-playing games as a vehicle for teaching history, culture and language. Sandbox Symposium Proceedings, July. Boston, MA

Kingston R, Carver S, Evans A, Turton I (2000) Web-based public participation geographical information systems: an aid to local environmental decision-making. Comput Environ Urban Syst 24:109–125

Lemke JL (1998) Metamedia literacy: transforming meanings and media. In: Reinking D, Labbo L, McKenna M, Kiefer R (eds) Handbook of literacy and technology: transformations in a post-typographic world. Lawrence Erlbaum Associates, Hillsdale, NJ, pp 283–301

Mehta M (2010) Dharavi redevelopment project. World conference—India, remaking sustainable cities in the vertical age, CTBUH 2010 conference, Mumbai, India, February 2010, presentation slides

McKelvey RD, Palfrey TR (1995) Quantal response equilibria for normal form games. Games Econ Behav 10(1):6–38

Nagel R (1995) Unraveling in guessing games: an experimental study. Am Econ Rev 85(5):1313–1326

Naquin CE, Kurtzberg TR, Belkin LY (2010) The finer points of lying online: E-mail versus pen and paper. Journal of Applied Psychology 95:387–394

NSDO n.a. National slum dwellers organization, Web. www.sdinet.org. Accessed 13 Jan 2016

Pickles J (1995) Representations in an electronic age: geography, GIS, and democracy. In: Pickles J (ed) Ground truth: the social implications of geographic information systems. Guilford Press, New York, pp 1–30

Poplin A (2015) How user-friendly are online interactive maps? Survey based on experiments with heterogeneous users. Cartogr Geogr Inf Sci 42(4):358–376. https://doi.org/10.1080/15230406.2014.991427. Taylor & Francis, ISSN: 1545-0465

Poplin A (2014) Digital serious game for urban planning: B3—design your marketplace! Environ Plann B Plann Des 41(3):493–511

Poplin A (2012a) Web-based PPGIS for Wilhelmsburg, Germany: an integration of interactive GIS-based maps with an online questionnaire. Special Issue J Urban Regional Inf Syst Assoc (URISA) 25(2):71–84

Poplin A (2012b) Playful public participation in urban planning: a case study for online serious games. Comput Environ Urban Syst (CEUS) 36(3):195–206. Elsevier, Web. http://www.sciencedirect.com/science/article/pii/S0198971511001116

Putnam LL, Jones TS (1982) The Role of Communication in Bargaining, Human Communication Research 8(3):262–280

Rinner C (1999) Argumentation maps—GIS-based discussion support for online planning. GMD research series No. 22. GMD German National Research Center for Information Technology, Sankt Augustin, Germany

Rinner C (2001) Argumentation maps—GIS-based discussion support for online planning. Environ Plann B Plann Des 28(6):847–863

Rinner C (2005) Computer support for discussions in spatial planning. In: Campagna M (ed) GIS for sustainable development. Taylor and Francis, London, pp 16–80

Rinner C (2006) Argumentation mapping in collaborative spatial decision making. In: Balram S, Dragicevic S (eds) Collaborative geographic information systems. Idea Group, Hershey, PA, pp 85–102

Ritterfeld U, Cody M, Vorderer P (eds) (2009) Serious games: mechanisms and effects. Routledge, New York

Roseland M (2005) Toward sustainable communities: resources for citizens and their governments. New Society Publishers, Gabriola Island, Canada

Schlossberg M, Shuford E (2005) Delineating "Public" and "Participation" in PP GIS. URISA Journal 16(2):16–26

Schlieder C, Kiefer P, Matyas S (2006) Geogames—designing location-based games from classic board games. IEEE Intell Syst 21(5):40–46

Schroeder P (1996) Report on Public Participation GIS Workshop, NCGIA Technical Report 96–97, Scientific Report for Initiative 19 Specialist Meeting

Sieber R (2006) Public participation geographic information systems: a literature review and framework. Ann Assoc Am Geogr 96(3):491–507

Sokolova M, Lapalme G (2012) How much do we say? Using informativeness of negotiation text records for early prediction of negotiation outcomes, Group Decision and Negotiation, Springer Netherlands, p. 363–379

Steinmann R, Krek A, Blaschke T (2004) Can online map-based applications improve citizen participation? Lecture notes in computer science, TED on e-government 2004. Springer Verlag, Bozen, Italy

Thompson MM (2000) GIS technology and data sharing, planning into the next millennium. Cornell J Plann Urban Issues 15:20–33

Zeng G, Takatsuka S (2009) Text-based peer–peer collaborative dialogue in acomputer-mediated learning environment in the EFL context, System 37, p. 434–446, Elsevier, Ltd.

Chapter 5
Addressing Uneven Participation Patterns in VGI Through Gamification Mechanisms

Vyron Antoniou and Christoph Schlieder

5.1 Introduction

The internet era since the turn of the century has been characterized by the ubiquity of open source, user generated data. The classic example of this "Web 2.0" paradigm is Wikipedia, an enormous, free, user generated encyclopedia. The geospatial domain entered the Web 2.0 era by leveraging user generated *spatial* content, known as *Volunteered Geographic Information* (VGI) (Goodchild 2007). Created by neo-geographers (Turner 2006), the most successful example so far is OpenStreetMap (OSM). OSM has motivated a constantly growing pool of contributors to create a free map of the world, providing up-to-date Geographic Information (GI) that was previously unavailable (Estes and Mooneyhan 1994; Goodchild 2007).

While OSM is quickly building a world map from open and freely available data, challenging points in the process have. One of the most discussed issues in the literature is the VGI data quality and fitness-for-purpose, often using OSM data as a paradigm (see for example Haklay et al. 2010; Jackson et al. 2013; Arsanjani et al. 2013; Kalantari and La 2015; Stein et al. 2015). Another closely related issue is that new sources of error and uncertainty related to social phenomena directly affect VGI datasets. These sources of error and uncertainty are fundamentally different from those of traditional authoritative datasets. This is mainly due to the social aspect of the VGI phenomenon and the crowd-based mechanism for data collection (see Antoniou and Skopeliti 2015 for an overview of the social factors that can affect VGI quality and participation patterns). For example, empirical research

V. Antoniou (✉)
Hellenic Army General Staff/Geographic Directorate, Athens, Greece
e-mail: v.anoniou@ucl.ac.uk

C. Schlieder
Faculty of Information Systems and Applied Computer Sciences, University of Bamberg, Kapuzinerstraße 16, 96047 Bamberg, Germany

© Springer International Publishing Switzerland 2018
O. Ahlqvist, C. Schlieder (eds.), *Geogames and Geoplay*, Advances in Geographic Information Science, https://doi.org/10.1007/978-3-319-22774-0_5

results show that contributors are affected by the underlying socioeconomic context of their activity area. There is a strong correlation between the Index of Multiple Deprivation for UK and the OSM completeness (Haklay 2010) and OSM positional accuracy (Antoniou 2011). Similarly, Girres and Touya (2010) note that socioeconomic factors (e.g. high income and low population age) result in a higher number of contributions. Zielstra and Zipf (2010) show that demographic factors such as the low population density areas (i.e. rural areas) have a direct impact on the completeness of VGI data. As the evidence converges on the question of how crowds behave, academic research has shifted focus to theoretical analyses of contributors: their nature, and motivation (Goodchild 2007; Coleman et al. 2009) and their new role in the production (Budhathoki et al. 2008).

Interestingly enough, the aforementioned observations hold true and similar patterns emerge for other sources of VGI, such as the popular photo-sharing web application of Flickr[1] that provides geo-tagged images. For example, Antoniou et al. (2010) show that by conducting a density analysis of geo-tagged images in Flickr, there are certain areas that attract more contributors than others resulting in the uneven distribution of user contributed data. However, the same type of spatial analysis conducted using the geo-tagged images of Geograph[2] found different results. Geograph is a spatially explicit application that implements a gamification approach to user contributed data, and provides more evenly distributed area coverage despite having fewer photos compared to Flickr (7993 Flickr photos vs. 1109 Geograph photos) (see Fig. 5.1).

Following this line of research, this chapter investigates the participation patterns of OSM contributors, the results produced, and the resulting impact on spatial data quality. The chapter provides solutions for counter-balancing unwanted effects of participation patterns through geogames. More specifically, in Sect. 5.2 we identify three issues of OSM mapping: (1) high productive contributors show little commitment to return and update geographic features they created, (2) the gap between the accumulated percentage of created features and the accumulated percentage of updated features is widening, (3) there is a significant contrast between areas of high and low mapping activity. Section 5.3 describes spatial allocation games as a subclass of location-based games suitable for addressing the participation issues. Based on an analysis of the geogames **Geograph, Foursquare, Ingress**, and **Neocartographer** six common design patterns for the allocation and deallocation of places are identified. We also show how the participation issues map onto the game design patterns. Finally, Sect. 5.4 describes the results from an agent-based spatial simulation that provide insights into the game flow of spatial allocation games. We present a model that distinguishes between a phase of fast and slow gameplay. Game designer should try to avoid the slow phase. The chapter concludes with a discussion of the results and an outlook on questions for future research in Sect. 5.5.

---

[1] www.flickr.com.

[2] www.geograph.co.uk.

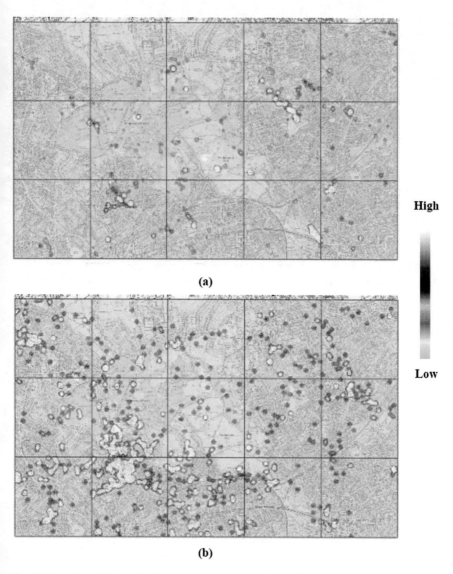

Fig. 5.1 Density surfaces for (**a**) Flickr and (**b**) Geograph in a test area of 3 × 5 km in North London. Source: Antoniou et al. (2010)

5.2 The OpenStreetMap Case Study

The focus area is the Greater London Area in UK, as the birthplace of OSM is University College London. Moreover, urban areas attract more OSM contributors and thus such areas facilitate the monitoring and the analysis of their digital

Table 5.1 Descriptive
statistics of the dataset

OSM data set: Greater London Area, UK	
Number of features	438,980
Number of unique contributing users	3230
Number of versions	917,000
Versions per feature (average)	2.09
Versions per user (average)	283.9

behavior. Instead of a direct bulk download from the OSM database, the dataset of
the area was downloaded in a shapefile format (by www.geofabrik.de) and then the
OSM API was used to collect only the necessary data for the analysis (changesets,
timestamps etc.) using the unique osmid of each feature. As in most wiki-based
projects, in OSM it is possible to trace back the lineage of each spatial feature and
thus monitor all the changes that each feature has undergone from its initial capture
up to date. The dataset contains 438,980 features that have in total 917,000 versions,
contributed by 3230 OSM contributors that have been active in the area since the
beginning of OSM. The descriptive statistics of the dataset are shown in Table 5.1.

   Although descriptive statistics give a basic understanding of the data at hand,
they do not shed light to the geographic distribution of the data or to any underlying
patterns. Thus, a more thorough analysis of the OSM datasets was conducted aiming to provide answers to the following questions relevant to the participation patterns and biases in OSM. Unwanted participation patterns and biases will be later
targeted by the gamification processes so to counter-balance them and thus enable a
more balanced volunteered contribution.

5.2.1   Is There a Commitment Between OSM Contributors and Their Spatial Edits?

From the early days of VGI, many scholars (see for example Elwood 2008;
Goodchild 2008; Heipke 2010) have recognized that one of the most important elements of crowdsourced spatial content is its unique relationship with local knowledge and its ability to capture such information. Indeed, in many crowdsourced
projects (see for example Haklay et al. 2010) local knowledge plays a central role.
However, when it comes to projects that extend to global scale, despite the numerous contributors that might be involved, the picture is quite different. An analysis of
the behavior of the most productive contributors (i.e. those that have contributed
more than 100 times) took place in order to examine how committed OSM contributors are to their edits, and thus see if local knowledge is present in such datasets. The
results showed that there is limited relationship between contributors and spatial
features. In fact, just a mere 10% of these high productive contributors are returning
to more than 20% of the features that have created in the past. This observation is
creating a fresh line of questions regarding the notion that contributors are bringing

along their local knowledge and the importance of this knowledge in the quality of the OSM dataset.

5.2.2 Is the OSM Dataset Kept Up-to-date by the OSM Contributors?

From the outset of OSM in 2004, the Greater London Area has been a very active area in terms of data crowdsourcing either through OSM mapping parties (i.e. leisure activities devoted to mapping an area) or through individual contributions. As a result, a highly detailed map of this area was available even from the first years of OSM. However, as more and more spatial features are portrayed in the OSM map, the contributors devote more of their energy to editing existing features than capturing new ones (Fig. 5.2). More specifically, until the second half of 2007 there were more additions of new features than updates of existing ones. However, from that point onwards the contributors focus more on updating OSM feature than creating new ones. This is a seemingly healthy evolution of OSM as in a densely built area the number of new features created is relatively small, while the editing of new ones could theoretically correct existing geometric and attribute errors, provide more detailed description of spatial features and improve the overall spatial data quality. Of course, this assumption holds true only if the editing effort is evenly distributed by OSM contributors to the entire population of spatial features available. In Sect. 5.2.3 we show that an uneven geographic distribution of edits has been observed and thus participation biases emerge.

Moreover, the examination of the edits' distribution shows a pattern where most of the features have only few edits while a small part of them gathers a large number of edits. More specifically, almost 66% of the features have up to 10 edits (note that the first edit is the creation of the feature) when at the same time about 12% of the features have more than 100 edits (Fig. 5.3). As explained below, interesting patterns also emerge when examining how up-to-date is the OSM dataset or how participation patterns are affected by the geography of the area.

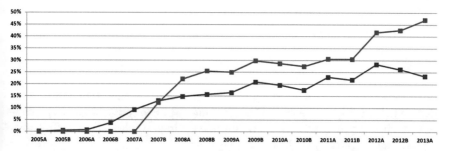

Fig. 5.2 The percentage of the active OSM contributors (out of the 3230 in total recorded) in Greater London Area that have created (*blue line*) and edited (i.e. updated) (*red line*) features

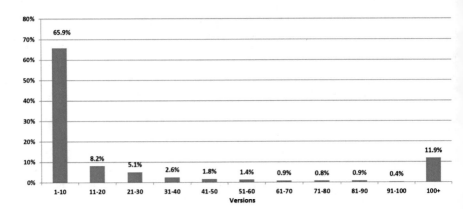

Fig. 5.3 Percentage of features per number of versions

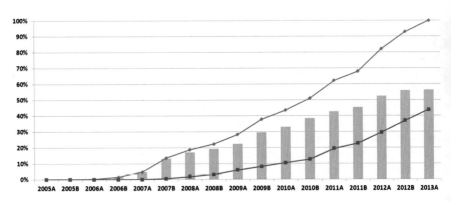

Fig. 5.4 The accumulated % of created features (*blue line*) vs the accumulated % of updated features (*red line*) by 6-month period. The *green bars* show the % of difference in the number of features

   In order to examine how up-to-date the OSM dataset is, the number of features created and edited in each 6-month period since the beginning of OSM has been calculated. There is a steadily growing difference (Fig. 5.4). More specifically, in the first half of 2013, over half (56%) of the geographic entries had not been updated/edited. This observation leads to the concerning conclusion that a growing number of OSM features are falling behind when the up-to-date factor is considered. This is even more interesting as the analysis showed that more contribution effort is focused on data updating than data creation (see Fig. 5.2) and thus this effort is not evenly distributed to the spatial features. Also, it is interesting to consider that, as previous research has shown, edits from many contributors improve the overall quality of the spatial features (Haklay et al. 2010).

5.2.3 Is There Any Spatial Pattern in the OSM Contributors' Behavior?

Spatial statistical analysis was completed (Hot Spot Analysis using the Getis-Ord Gi* statistic provided in ESRI ArcGIS 10.2.2) on the dataset using the number of edits of each OSM feature. The null hypothesis is that all areas equally attract contributors' interest in terms of editing existing features. Any deviation from this hypothesis should be a random phenomenon (an exception could be in the cases of major reconstruction works in specific areas that could lead to increased number of edits, however no such thing has been recorded for the study period). It was possible to identify statistically significant spatial clusters of both high values (hot spots) and low values (cold spots) in the number of edits for each feature. The Hot Spot Analysis based on the number of edits of each road segment reveals which areas are attracting the interest of OSM contributors and which are not and thus challenges the validity of the null hypothesis. Figure 5.5 shows the results of the hot spot analysis of the streets of the Camden London Borough. It is interesting to note that the hot areas are the area around UCL (lower red) and the famous and highly touristic are of Camden Market (middle red). This observation shows that users are focusing their contribution on specific popular and well-known areas while overlooking other more obscure ones and thus the null hypothesis is rejected. The results of this analysis are in accordance with previous observations about the correlation of socio-economic factors and contribution mentioned above.

The analysis of the OSM dataset shows that the evolution of VGI introduced new uncertainty sources for the spatial data available on the Web. Apart from the classic spatial quality elements (ISO 2005), there is growing evidence that social factors influence data quality of VGI. It is worth noting that these social factors are totally different from the error sources that usually affect the classic geographic information production mechanisms followed by national mapping agencies. Consequently, the VGI opens up new areas for further research in the subject mat-

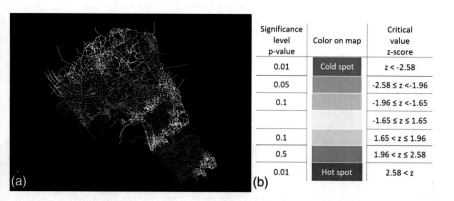

Fig. 5.5 (a) OSM participation pattern based on the hot spot analysis on the versions of each feature, (b) the z-score and p-value distributions (source: ESRI)

ter of spatial data quality and the long-run evolution of initiatives like OSM. Measures to counter-balance social data phenomena should be taken into account if OSM is to become a world-class spatial database, and/or simply preserve its current status as a reliable, up-to-date spatial database.

5.3   Spatial Gamification

One solution to the data issues emerging from VGI, such as those identified for OSM in the prior section, is using techniques used in geogames to increase updated data from less popular places. Crowd-based mapping shares some characteristics with the game playing activities in geographic space that have become popular, such as Niantic's global multi-player game **Ingress** (Hodson 2012) or the gamification mechanisms of the location-based social network **Foursquare** (Lindqvist et al. 2011)—not to mention the current ecstasy about **Pokémon Go**. In such location-based geogames, the geographic location of the player constitutes a fundamental game element since different places in the geographic environment are associated with different choices of game actions (Schlieder et al. 2006). In the following, we use the term geogame to refer to such location-based games, thus, in a more restricted sense than in the introductory chapter of this book. Geogames motivate players to visit places which they probably would not visit outside the game. Not surprisingly, researchers have started to study geogames as a means to increase participation in VGI. Examples include the photographing and geo-referencing of buildings (Matyas et al. 2009) or the mapping of noise in an urban environment (Garcia-Martí et al. 2013). However, little is known on how to relate the participation issues of VGI mapping to specific design patterns of geogames.

The following analysis concentrates on the design of the game mechanics, that is, the set of rules which define the sequence of game actions (Montola et al. 2009; Adams and Dormans 2012). Geogames are played as search games or chase games or they follow some other paradigm, frequently, a paradigm drawn from pre-computer outdoor games (Davidsson et al. 2004). The many variants of capture-the-flag games constitute such a time-tested paradigm. Places in the geographic environment act as resources that the mechanics of the geogame allocate to the players according to a variety of rule sets. This analysis refers to such games as spatial allocation games.

Geograph, the photographic mapping activity mentioned in the chapter's introduction, can be seen as a spatial allocation game in which the player's task consists in submitting the first geographically representative photograph for squares of the Ordnance Survey grid of Great Britain and Ireland[3]. Three other geogames illustrate other ways to interpret the spatial allocation paradigm. **Foursquare** primarily offers the services of a location-based social network, however, the check-in mechanism implemented in releases prior to Foursquare 8.0 adds a gamification element

---

[3] The full name of the game is Geograph Britain and Ireland (www.geograph.org.uk).

(foursquare.com). Users who check-in at places with a mobile device are rewarded for frequent re-visits by becoming the "Mayor" of the place. **Ingress** is a geogame in which two teams of players compete to capture and re-capture places called "portals" on a global game board (ingress.com). The game comes with complex game mechanics which—to simplify considerably—allocates a portal to the team of the player who visits the place while being in possession of the appropriate game resources and knowing how to best deploy them tactically. **Neocartographer** has been designed by the second author of this article as a game for two competing players or teams. The players try to obtain places which form a particular spatial configuration, instead of just maximizing the number of places in their possession (www.geogames-team.org).

5.3.1 Design Pattern for Allocation Games

Some fundamental design choices apply to all types of geogames as pointed out by Montola et al. (2009). Geogames are either played on a bounded game field or without spatial restrictions anywhere in the global geographic space. In the temporal dimension, the game can last for a limited playing time or can go on without end (pervasive play).

The specific design choices for spatial allocation games have not been systematically described in the literature. For the purpose of this analysis, two design parameters are considered: allocation type and place-to-player ratio. In three of the four example games, place allocation is exclusive, that is, a place can only be allocated to one player or team at a time. **Foursquare**, where several users can check-in at the same place and earn badges for these check-ins, uses multiple allocation in combination with exclusive allocation for awarding the title of mayor of a place.

A simple metric of ratio of places to players reveals further differences. **Geograph** is played by 12,038 players (accounts) in 331,956 places (grid cells)[4], reflecting a place-to-player ratio of approximately 30:1. In contrast, the **Foursquare** website states for the same period that more than 1,500,000 places (venues) created by businesses are visited by more than 45,000,000 players (patrons) which amounts to a ratio of 1:30. **Ingress** does not publish global player statistics. However, since only two teams compete, the ratio is of the same order of magnitude as the global number of portals. A typical **Neocartographer** game where two players compete for half an hour is played with ten places, that is, with a place-to-player ratio of 5:1 (Table 5.2).

Most geogames with a large place-to-player ratio are based on a mechanics with exclusive place allocation. In OSM mapping, the geographic features outnumber the mappers by far with a place-to-player ratio even higher than that of **Geograph** or **Ingress**. A global and pervasive game play with exclusive allocation suggests itself as design choice for a gamification approach to OSM mapping. Different game design patterns for allocating and deallocating places are consistent with this choice.

---

[4] As of April, 2014.

Table 5.2 Design parameters of spatial allocation games

	Spatial boundary	Temporal boundary	Allocation type	Place-to-player ratio r
Geograph	Game field	Pervasive play	Exclusive	$10 < r < 100$
Foursquare	Global	Pervasive play	Multiple	$10^{-2} < r < 10^{-1}$
Ingress	Global	Pervasive play	Exclusive	$10^5 < r < 10^6$
Neocartographer	Game field	Playing time	Exclusive	$1 < r < 10$

Table 5.3 Design patterns for allocating places

	Mechanics	Objective	Example
First-to-visit	The place goes to the first visitor	Spatial coverage	Geograph points (Geograph) Claiming a portal (Ingress) Claiming a cell (Neocartographer)
Nth-to-visit	The place goes to the n-th visitor	Game balancing	Second visitor points (Geograph)
Most-revisits	The place goes to the most frequent visitor	Revisit frequency	Mayor of a place (Foursquare)

An allocation pattern describes the game mechanics which specify what players need to do in order to obtain a place. A widely used mechanism consists in allocating the place to the first visiting player (first-to-visit pattern, Table 5.3). In a later stage of the game, when most of the places have been allocated to their first visitors, some games employ an additional mechanism which awards the place also to the second or even third visitor in a multiple allocation scheme (nth-to-visit **pattern**, Table 5.3). Geogames with a small place-to-player ratio tend to reward players who revisit a place (most-revisits pattern, Table 5.3).

Deallocation mechanisms counteract the consumption of places by allocation mechanisms. Some games, such as the original version of **Geograph**, do not use deallocation at all (never pattern, Table 5.4). The most popular mechanism for competitive two player games permits the player to reclaim a place from the opponent. In the simplest form, a place is reallocated any time one of the two players visit it (when-reclaimed pattern, Table 5.4). Another solution consists in using a decay time after which places are freed (when-decayed pattern, Table 5.4).

Although the aforementioned design patterns for allocation and deallocation do not provide an exhaustive inventory of design choices, they help to identify possible gamification approaches to the quality issues of OSM mapping. The design objective of maximizing the revisit frequency matches the problem that OSM contributors show little commitment to their edits (most-revisits pattern, Table 5.3). Similarly, the when-reclaimed pattern for deallocation motivates players to visit places which other players have visited before. It permits to address the design

Table 5.4 Design patterns for deallocating places

	Mechanics	Objective	Example
Never	The place is allocated for the whole game	Simplicity	Geograph points (Geograph)
When-claimed	The allocation changes if another player meets the allocation criterion	Data recency game balancing	Reclaiming portals (Ingress)
When-decayed	After a time span, the allocation is cleared	Game balancing	Energy loss of resonators (Ingress) Moving time window (Foursquare) Time-gap points (Geograph)

objective of data recency in a VGI game context. The issue of spatial regularities in the behavior of OSM contributors seems more complex as it reflects the effects of socio-economic factors. Gamification might still counteract the spatial clustering of mapping activities. The first-to-visit pattern combined with exclusive allocation is a well-tried mechanism for maximizing the spatial coverage of in-game activities.

5.3.2   Place Allocation and Game Flow

Allocation measures the accumulated percentage of places that have been assigned to players as a function of the percentage of total playing time. It runs from 0% allocated places at 0% time to some percentage smaller or equal 100% allocation at the end of the game (100% time). This simple measure captures an important aspect of the game flow, especially if changes in allocation are observed: a low allocation rate indicates a phase in which the game's goal is hard to achieve, while a high rate corresponds to a phase of easier game play.

In principle, a game can allocate places at a constant rate until no more places are available. It is not difficult to design game mechanics with that effect (Box 5.1). There are some advantages of a constant allocation rate, most importantly, the control over total playing time. In organizing a gamified mapping event, which targets a specific feature class and spatial region, for instance, parking lots on a university campus, it helps being able to anticipate the progress of the game. Furthermore, constant allocation implies constant consumption of additional resources that the game might require, such as the verification of the mapped features.

Most game mechanics do not guarantee a constant allocation rate in all games, although some display a similar behavior on average. In general, the allocation rate of a game changes over time and depends on a number of factors such as the spatial distribution of players or the type of locomotion they use. Allocation is empirically determined by logging the game and expressed as a real-valued function $a(t)$: $[0,1] \rightarrow [0,1]$. Figure 5.6a shows a typical plot of allocation as a function of time.

Box 5.1 Game Mechanics with Constant Allocation Rate

The task of the players consists to map geo-features at 1000 distinct places. Each day a maximum of ten places are allocated to players that have mapped the corresponding features. The game client informs the players in real-time on the number of places that can still be mapped that day. After the daily upper bound is reached, the game server ignores all allocation requests. There is no queuing of requests. If at the end of the day less than ten places have been mapped, a place lottery allocates the remaining places to some randomly chosen players. No geo-features are mapped in places that have entered the lottery.

The game mechanics produces an allocation rate of exactly ten places/day and ends the game after a period of exactly 100 days. To minimize or avoid the place lottery, the daily maximum should be sufficiently small. It is also a good idea to try to estimate the difficulty of mapping a place and put "easy" places into the lottery. Since the difficult mapping tasks are accomplished during the game, chances are good that the few remaining places will be mapped without the incentive provided by the game.

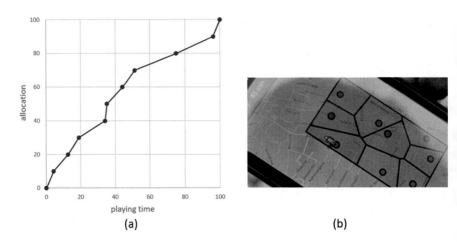

Fig. 5.6 Neocartographer (**a**) allocation plot, (**b**) interface

The values on both axes are expressed as percentages and describe ten allocation events of a **Neocartographer** game. The additional data point (0,0) has been added to describe the status at the start of the game. This specific instance of the game has been played on a rather small and sparsely populated (n = 10 places) game field in the city of Augsburg, Germany. Two players were competing for the allocation of the places. The game play followed the mechanics described in Box 5.2. Figure 5.6b shows the interface of the client software at the end of a (different) game. It is important to realize that the allocation of a place constitutes an event. Allocation

Box 5.2 Neocartographer
In its simplest version, Neocartographer is a two-player game in which each player tries to be the first to accomplish specific mapping tasks at the n places shown by the game client software. From the point of view of allocation, the game mechanic is very simple. Whoever maps a place first, owns that place, a straightforward application of the first-to visit pattern with no deallocation.

The main point of the game mechanic, however, is to challenge the players to reason spatially. Places are not equally interesting to own. The worth of a place is determined by the size of its Voronoi cell. In the standard playing mode, the boundary of the cell is not shown to the players. They have to decide on which place to move to next taking into account the location of the opponent, the distance of the place, and what they believe, the size of its cell is.

Game logs of the two-player version of Neocartographer show that for small game fields with uniformly distributed places, the game mechanic produces a relatively constant allocation rate on average.

changes in a stepwise way. At 51% playing time, allocation increases to 70% and at 60% playing time it still has the same value (Fig. 5.6a). The line segments interpolating between the data points are shown only to illustrate the concept of allocation rate, with a steeper slope of a segment indicating a higher rate.

Allocation is a discontinuous step function $a(t)$ if a geogame is played on a finite number of places. For the purpose of visualization and analysis, however, the data points of $a(t)$ are often interpolated to produce a continuous approximation of the allocation function. If no deallocation of places occurs, the allocation function is monotonic: for any two time points u and v during playing time, $u < v$ implies $a(u) \leq a(v)$. Many monotonic allocation games are played until all places have been allocated, that is, $a(1) = 1$. We call them *simple geogames*. A simple geogame has a *convex allocation rate* if for any two time points u and v during playing time, the graph of the allocation function $a(t)$ lies below or on the line segment through $a(u)$ and $a(v)$ (Fig. 5.7a). If the graph lies above or on the line segment for all choices of u and v then the allocation rate is called *concave* (Fig. 5.7b). Convex and concave are not mutually exclusive properties: constant allocation is both convex and concave. As the example from Fig. 5.7c illustrates, an allocation function may also be neither convex nor concave.

In convex geogames, place allocation starts slow and accelerates towards the end of the game. Box 5.3 describes a game mechanic with convex allocation. Many designers consider it a flaw if a game becomes progressively easier to play. They fear that such a game scares off novice players by its difficulty and bores expert players by not providing sufficient challenge during the later phases of play. In fact, none of the four geogames considered in this chapter are designed to produce convex allocation.

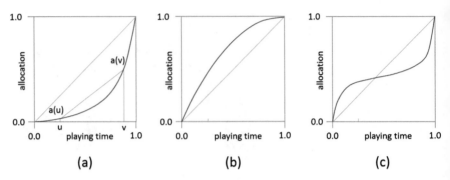

Fig. 5.7 Allocation function of simple geogames (**a**) convex, (**b**) concave, (**c**) neither convex nor concave

---

Box 5.3 Game Mechanics with Convex Allocation Rate

The task of the players consists of mapping geographic features at 1023 distinct places. At the beginning of the game, a single geographic location is disclosed to the players. This place will be easy to reach for some players, but for most players it will be far from their current geographic position and thus difficult to reach. The place is allocated to the first player who moves there and accomplishes the mapping task. After 10% of the total playing time, two more places are disclosed, and after 20% another four places, and so on until the last disclosure occurs at 90% playing time. The last disclosure of places informs the players about the geographic position of the remaining 512 distinct places.

Places are disclosed according to a power law. Assuming a uniform random distribution of players and places, the chance to reach a place before other players doubles in each successive phase of the game. Allocation should increase from phase to phase provided that the player base maintains its activity level throughout the game.

Nevertheless, the idea of increasing the place allocation rate during game play makes sense for some spatial gamification scenarios. One reason for adopting convex allocation is to counterbalance the effects of physical fatigue in games that involve locomotion over long distances and long periods of time, especially when weather conditions are bad. Another reason is that the game mechanics may need to deal with places that differ considerably in the difficulty of the associated mapping task (e.g. harder to access, more time consuming). Convex game mechanics such as the one from Box 5.3 may be used to balance task difficulties by disclosing first the places that are the hardest to map leaving the easier cases for the end game.

5.4   Simulation Studies: The Problem of Accumulated Advantage

Most game mechanics do not follow convex allocation. They make it easy for players to obtain place allocations at the beginning of the game while making allocation progressively harder towards the end. A number of questions are relevant to the game designer interested in such a concave allocation: Under which assumptions does such an allocation arise? Is it possible to distinguish different phases in place allocation? How do the phases affect the game play? Simulation studies can help to address some of these questions.

Since most geogames do not permit the modification of their game mechanics (e.g. **Geograph, Ingress, Foursquare**), it is difficult to field-test the causal effects of allocation patterns. Additionally, field-testing geogames that use a large number of places requires a considerable time investment by large numbers of players. In response, some designers have resorted to integrating testing into playing. However, any optimization of the game mechanics means changing the rules during the game, which is unpopular among players, to say the least. An alternative approach, using game simulations has been successfully applied to the design of video games (Adams and Dormans 2012). One of the issues that simulation testing highlights is the problem of accumulated advantage, which is discussed more in this section.

One critical element of location-based game simulation is the model of spatial player behavior. Many degrees of realism are possible, including agent-based simulations that interact with a terrain model (Heinz and Schlieder 2015). In simulations, much less input is needed to gain a better understanding of place allocation. Qualitative abstractions that permit the comparison of results between different simulation runs and between modifications of the player model are also quite helpful. The following phases of concave allocation described in this section provide such an abstraction for comparison.

A very simple model of casual game play is used to illustrate this type of analysis. The software agents that model the players show alternating phases of spatial activity and inactivity. In an activity phase, the agent moves according to a constant velocity random walk direction mobility model (e.g. Roy 2011). This reflects the casual character of game play. The player does not actively travel towards the places that are of interest in the game, rather, the player uses journeys he or she engages in for some other reasons (e.g. commuting to work). A more elaborate model would take into account that players are sometimes willing to deviate from their routine journeys to gain some advantages in the game. Spatial inactivity corresponds to phases in which the agent does not change the location such as is the case for most types of work or leisure activity. For a casual player, the latter phases are typically longer than the phases devoted to locomotion.

The simulation uses a grid-based representation of the spatial environment. Places where the game invites players to map geographic features are represented as a single cell each. The simulation is implemented in the NetLogo 5.05 environment

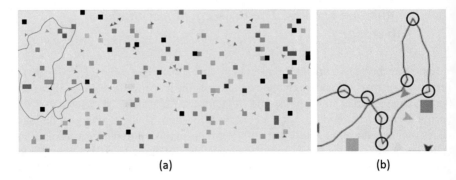

Fig. 5.8 Simulation (**a**) unallocated places (*black cells*), allocated places (*colored cells*), and players (*colored triangles*), (**b**) activity phases

(Wilensky 2012), which supports grid-based spatial simulations. Figure 5.8a shows a typical scenario: the agents (colored triangles) move through the grid along random walk trajectories (shown for one agent). The places at which the game invites to engage in geographic mapping activities are shown as colored cells. Cells that have been allocated to an agent are colored in the corresponding color. The black cells are still available for allocation. The game state is depicted at a later stage where all but 17 of the 128 places have been allocated to players.

Although Fig. 5.8a does not convey the dynamics of the simulation, the location of the phases of inactivity can be identified to a certain extent from the players' trajectory because of the specific random direction model. It lets the agents move along rather smooth paths. At brisk turns in the trajectory, the agent, very likely, has paused for a phase of inactivity. In Fig. 5.8b, which is from a different simulation run, the locations of inactivity have been marked by circles.

The simulation models exclusive allocation to the first visitor with no deallocation, this is similar to the game mechanics of the original version of **Geograph** or that of **Neocartographer**. Interestingly, a single form of allocation function consistently emerges in the simulation runs. Figure 5.9 shows the percentage of allocated places as a function of the percentage of total playing time for a typical simulation. In the example, 20% of the places are allocated in less than 5% of the playing time and to allocate 80% of the places, it needs less than 40% of the time. The same qualitative behavior is found in repeated simulation runs. As in the much smaller example from the **Neocartographer** game (Fig. 5.6a), the linear interpolation of the simulated allocation function is neither convex nor concave. Its overall shape, however, indicates that the game start with high allocation rates and then slows down considerably. In such a case, allocation can be approximated by a concave function.

Fitting a concave function to the simulation distinguishes two phases: a phase of fast play during which the allocation change is above average followed by a phase of slow play during which allocation change is below average (Fig. 5.10a, b). For obvious reasons, the slow play phase is not very attractive to human players. Players experience most success, in the sense that they are mapping features, during the first part of the fast play phase when there are more places than players.

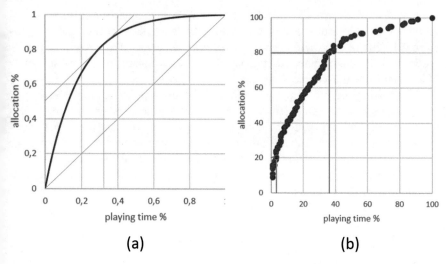

Fig. 5.9 (**a**) Simulated first-to-visit allocation with no deallocation. (**b**) Least-square fit of a concave function (logarithm)

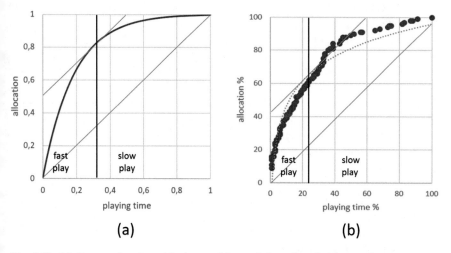

Fig. 5.10 (**a**) Concave function with phases of fast and slow play. (**b**) Phases of gameplay in a simulation run

The simulation reveals another problem of a game mechanics based on exclusive first-to-visit allocation with no deallocation: the outcome of the game is predictable at an early phase. From the top 10% highest scoring players at the moment when 50% of places are mapped, most will still be among the top 10% at the end of the game. An advantage early in the game accumulates with this game mechanics. This phenomenon is called the problem of accumulated advantage in game design.

To avoid the slow-down, first-to-visit allocation without deallocation should not be played beyond the point of 50% mapped places. Note that **Geograph** which implements this pattern, has already allocated 82% of the places to players. Based on the simulation results, one would predict that a spatial allocation game at this stage is mostly of interest to the highest performing players. This prediction is consistent with the high score lists published by the **Geograph** project which show little change over the course of the years.

Remediating the problem by deallocation is not straightforward. Two-player geogames often combine exclusive allocation to the first visitor with the when-reclaimed deallocation pattern. Simulations show that frequently a new type of problem with the game balance emerges: the outcome of the game becomes too unpredictable as it is virtually decided in the last cycles of the simulation. For the playing experience, advantages which do not accumulate at all are as frustrating as advantages which accumulate too fast.

5.5   Conclusions and Outlook

This chapter presented first results relating participation patterns in VGI to gamification mechanisms which can help to address participation issues. A case study from the Greater London Area revealed three spatial pattern in the behavior of OSM contributors. First, highly productive contributors were found to show little commitment to return and update the geographic features they created (commitment problem). Second, the gap between the accumulated percentage of created features and the accumulated percentage of updated features is widening (update problem). Third, the spatial analysis of OSM feature version shows a contrast between areas of high and low mapping activity (clustering problem).

The chapter described spatial allocation games as a class of geogames suitable for a gamification approach to approach these issues in VGI mapping. Two design choices specific to allocation games were identified, the allocation type and the place-to-player ratio. The analysis of the geogames **Geograph, Foursquare, Ingress**, and **Neocartographer** helped to specify six common design patterns for the allocation and deallocation of places, and mapped the VGI participation issues onto design patterns. The commitment problem can be addressed by the most-revisits allocation pattern, the update problem by the when-reclaimed deallocation pattern, and the clustering problem by the first-to-visit allocation pattern.

Agent-based simulations can help understanding the effect of allocation and deallocation patterns. The phase model based on the approximation of the allocation function helps to describe the two phases of fast and slow play produced by game mechanics based on concave allocation. To the best of our knowledge, simulations studies have not been systematically used to design geogames. Results from the first trials are encouraging. The simulation correctly reproduced the problem of accumulated advantage which arises in the Geographing game and links it to first-to-visit allocation pattern. As for any type of simulation, realism remains a challenge, especially when modeling player motivation. However, the costs for play testing geogames are much higher than for classical video games, which is why simulations offer an interesting alternative to study their game mechanics.

Future research will explore more complex game mechanics which go beyond the combination of a single positive feed-back loop (allocation) with a single negative feed-back loop (deallocation). Preliminary results show that agent-based simulation provides a valuable method for avoiding the repeated modifications of the game mechanics by trial and error, which geogames currently impose on their players.

Acknowledgments The second author's research has been funded in part by ESRI Inc. within the Geogames and playful Geodesign project.

References

Adams E, Dormans J (2012) Game mechanics: advanced game design. New Riders Games, Berkeley, ISBN 978-0-321-82027-3

Antoniou V (2011) User generated spatial content: an analysis of the phenomenon and its challenges for mapping agencies, Doctoral thesis [Online] http://discovery.ucl.ac.uk/1318053/. Accessed 18 March 2014

Antoniou V, Haklay M, Morley J (2010) Web 2.0 geotagged photos: assessing the spatial dimension of the phenomenon. Geomatica 64(1):99–110. Special Issue on VGI

Antoniou V, Skopeliti A (2015) Measures and indicators of VGI quality: an overview. ISPRS Ann Photogramm Remote Sens Spatial Inf Sci 1:345–351

Arsanjani JJ, Barron C, Nakillah M, Helbich M (2013) Assessing the quality of openstreetmap contributors together with their contributions. In: 16th AGILE international conference of geographic information science, Leuven, Belgium, pp 14–17

Budhathoki NR, Bruce B, Nedovic-Budic Z (2008) Reconceptualizing the role of the user of spatial data infrastructure. GeoJournal 72(3–4):149–160

Coleman DJ, Georgiadou Y, Labonte J (2009) Volunteered geographic information: the nature and motivation of producers. Int J Spatial Data Infrastruct Res 4:332–358

Elwood S (2008) Volunteered geographic information: key questions, concepts and methods to guide emerging research and practice. GeoJournal 72(3-4):133–135

Davidsson O, Peitz J, Björk S (2004) Game design patterns for mobile games. Project report. Nokia Research Center, Finland

Estes JE, Mooneyhan W (1994) Of maps and myths. Photogramm Eng Remote Sens 60(5):517–524

Garcia-Martí I, Rodríguez-Pupo L, Díaz L, Huerta J (2013) Noise battle: a gamified application for environmental noise monitoring in urban areas. In: Vandenbroucke D et al (eds) AGILE-13. Springer, New York. www.agile-online.org/index.php/conference/proceedings/proceedings-2013

Girres JF, Touya G (2010) Quality assessment of the French OpenStreetMap dataset. Trans GIS 14(4):435–459

Goodchild MF (2007) Citizens as sensors: the world of volunteered geography. GeoJournal 69(4):211–221

Goodchild MF (2008) Commentary: wither VGI? GeoJournal 72(3-4):239–244

Haklay M (2010) How good is volunteered geographical information? A comparative study of OpenStreetMap and Ordnance Survey datasets. Environ Plann B Plann Des 37(4):682–703

Haklay M, Basiouka S, Antoniou V, Ather A (2010) How many volunteers does it take to map an area well? The validity of Linus' law to volunteered geographic information. Cartogr J 47(4):315–322

Heipke C (2010) Crowdsourcing geospatial data. J Photogramm Remote Sens 65(6):550–557

Heinz T, Schlieder C (2015) An Agent-based simulation framework for location-based games. In: AGILE-15, proceedings of international conference on geographic information science, online proceedings. agile-online.org/index.php/conference/proceedings/proceedings-2015

Hodson H (2012) Google's Ingress game is a gold mine for augmented reality. New Scientist 216(2893):19. https://doi.org/10.1016/S0262-4079(12)63058-9

International Organisation for Standardisation (2005) 19113 geographic information—quality principles. ISO, Geneva

Jackson SP, Mullen W, Agouris P, Crooks A, Croitoru A, Stefanidis A (2013) Assessing completeness and spatial error of features in volunteered geographic information. ISPRS Int J Geo-Inf 2(2):507–530

Kalantari M, La V (2015) Assessing OpenStreetMap as an open property map. In: OpenStreetMap in GIScience. Springer International Publishing, Switzerland, pp 255–272

Lindqvist J, Cranshaw J, Wiese J, Hong J, Zimmerman J (2011) I'm the mayor of my house: examining why people use foursquare-a social-driven location sharing application. In: SIGCHI-11, Proc. conf. on human factors in computing systems, pp 2409–2418

Matyas S, Matyas C, Mitarai H, Kamata M, Kiefer P, Schlieder C (2009) Designing location-based mobile games: the CityExplorer case study. In: de Souza e Silva A (ed) Digital cityscapes. Merging digital and urban playspaces. Lang, New York, pp 187–203

Montola M, Stenros J, Waern A (eds) (2009) Pervasive games. Theory and design. Morgan Kaufmann, Burlington, MA

Roy RR (2011) Handbook of mobile ad hoc networks for mobility Models. Springer, New York

Stein K, Kremer D, Schlieder C (2015) Spatial collaboration networks of OpenStreetMap. In: Jokar Arsanjani J et al (eds) OpenStreetMap in GIScience. Springer, Netherlands, pp 167–186

Schlieder C, Kiefer P, Matyas S (2006) Geogames: designing location-based games from classic board games. IEEE Intell Syst 21(5):40–46

Turner AJ (2006) Introduction to neogeography. O'Reilly Media Inc., Sebastopol, CA

Wilensky U (2012) NetLogo 5.0. http://ccl.northwestern.edu/netlogo/. Center for Connected Learning and Computer-Based Modeling, Northwestern University. Evanston, IL

Zielstra D, Zipf A (2010) A comparative study of proprietary geodata and volunteered geographic information for Germany. In: Proceedings of the thirteenth AGILE international conference on geographic information science, Guimarães, Portugal

Chapter 6
Teaching Geogame Design: Game Relocation as a Spatial Analysis Task

Christoph Schlieder, Dominik Kremer, and Thomas Heinz

6.1 Introduction

Before we can play games, we have to design them. While both activities are complementary, educational uses of games have focused on the playing aspect. There are, however, a number of reasons for including game design into a curriculum on spatial thinking.

First, location-based games are known to be quite effective at supporting a broad variety of learning processes (Klopfer 2008; Schaal et al. 2012). In education practice, however, we see little variation in the underlying game mechanics. A study of mobile location-based learning projects in Germany, for instance, revealed that all projects, which used gamification approaches, were adopting some variant of **Geocaching** (Lude et al. 2013).

A second reason is that location-based games have started to become part of the media and entertainment environment of teenagers. Students in grade 10 or higher very likely have had contact with non-educational versions of these games either as players or as viewers of Let's Play videos. **Ingress**, operated by Niantic, is one of the early popular examples of a location-based game (Hodson 2012). Researchers have also begun to look at **Pokémon Go**, an enormously successful game released by the same company in 2016 (Althoff et al. 2016). Furthermore, commercial gamification approaches increasingly use geographic location, for instance, in the form of location-based games that are marketing a touristic destination (Celtek 2010).

C. Schlieder (✉)
Faculty of Information Systems and Applied Computer Sciences, University of Bamberg,
Kapuzinerstraße 16, 96047 Bamberg, Germany
e-mail: christoph.schlieder@uni-bamberg.de

D. Kremer • T. Heinz
University of Bamberg, Bamberg, Germany

© Springer International Publishing Switzerland 2018
O. Ahlqvist, C. Schlieder (eds.), *Geogames and Geoplay*, Advances in
Geographic Information Science, https://doi.org/10.1007/978-3-319-22774-0_6

Finally, the design process of a location-based game involves a considerable amount of spatial analysis. When game designers identify suitable places for game play, they compare, for instance, the distances between the places based on assumptions about pedestrian locomotion. Such GIS supported analysis tasks fit very well into a curriculum on spatial thinking—see Sinton and Lund (2007) for examples of teaching spatial thinking in a multi-disciplinary context.

However, the design of a location-based game requires a basic understanding of spatial game mechanics, software engineering skills, considerable knowledge of the geography of the place, which the players are going to explore, and, last, but not least, a sufficient amount of time for creating, testing, and improving the design. From a teaching perspective, these requirements seem prohibitive, even in secondary education. This chapter approaches the challenge of creating geogames as a classroom activity from the perspectives of location-based game design, spatial analysis and geo-information processing. We describe an approach, which avoids much of the complexity of the design process by applying *three heuristics*: (1) We start with a rule set known to produce a well-balanced game and do not ask the students to invent the rules of the games. (2) We proceed with visiting the geographic environment, which acts as the game field and do not let the students design the game based on web cartography only. (3) We challenge the students to optimize the game flow by applying spatial analysis and do not encourage chance results.

We will support the heuristic principles outlined above in this chapter and present the methodological and technical tools to implement them. Section 6.2 reviews related work and presents a subclass of location-based games, which has been studied extensively in the research literature. Referring to a model of the game design process, we show in Sect. 6.3 how to reduce the complexity of the design task by focusing on game relocation, that is, the task of adapting a successful location-based game to a new environment. Relocating a geogame involves redesigning the places of gameplay. Section 6.4 illustrates how place design operates using the geogame **CityPoker** as an example. We adapt the place design process to our learning scenario by decomposing it into two phases and devote a separate section to each of them. Section 6.5 describes the place storming method, which helps students to identify places in the geographic environment that are suitable for game actions. In Sect. 6.6, we present a software tool that supports the spatial analysis tasks involved in the design of a location-based game. This tool assists students to relocate a geogame in a classroom project. Finally, we conclude in Sect. 6.7 with a discussion of the results we have obtained as well as of future research directions.

6.2 Related Work and Basic Terminology

In this chapter, the term *geogame* refers to competitive location-based games in which at least two players or two teams of players move in an urban or natural environment using mobile devices to communicate, to access spatial data, and to solve place-related tasks (Schlieder et al. 2006). As a characteristic design feature, such geogames allow players to perform game actions only at certain geographic locations. We call them *places of game play* (POG). In a competitive two-player variant of **Geocaching**, for instance, the geocaches act as the POG. They are the only places where the player can perform the game action of finding a hidden object by physically accessing the cache. All geogames create some sort of virtual game board, which adds a layer of semantics to the geographic environment.

There are different ways to exploit the additional layer of spatial semantics. A live action role playing game creates POGs such as a magic forest, which may have little resemblance with features of the geographic environment (Montola et al. 2009). Other genres of geogames adopt a similar approach. **Ingress**, for instance, advertises itself with the promise: "The world around you is not what it seems". In serious games, however, designers often do not aim at creating an alternate reality distinct from geographic reality (De Gloria et al. 2012). Instead, they try to link the players by spatial actions to existing places in the environment (von Borries et al. 2007; de Souza 2009; Schaal et al. 2012). Depending on the geogame's objective— a learning outcome, a tourist experience—the designer chooses different *places of interest* (POI) to act as the POGs of the game. The choice of places has to be made for any geographic region where the game is played. We discuss the challenges of this game relocation task in Sect. 6.3.

In practical terms, game relocation often involves outdoor exploration. First-hand experience helps to determine which places support which game activities. Kristiansen et al. (2014) describe site storming, an outdoor exploration method for discovering POG and associated game tasks. Our simplified method, which we call place storming, provides more guidance to the students than site storming and it supports finding typical combinations of places and activities rather than exceptional ones. In our approach to game design in the classroom, the students engage first in a phase where they use place storming to generate a set of POI. This set consists of places the students identified as interesting because the players could engage in actions that contribute to the learning outcomes (Fig. 6.1).

Place storming generates more places than are actually needed for the game. The surplus is important since not every subset of POI qualifies as POG. Additional constraints have to be considered during the design phase of the geogame. Important constraints arise from the balancing of the game mechanics, that is, from the way in which the rules of the geogame interact with the places. Balancing aims at rules that make the game neither too easy nor too hard to play. Another objective of balancing is fairness. The choice of POG should not give an advantage to one of the teams. Game design research has described a number of balancing methods outside the realm of location-based games (Adams and Dormans 2012). For the case of location-based games, Schlieder et al. (2006) have shown that the game designer has to provide some mechanism to compensate for the differences in locomotion speed, which always exist between players. Otherwise, a trivial winning strategy dominates: the fastest player nearly always wins. Because of the importance of locomotion, the time that players need to move between the POG becomes the most important parameter to analyze in the context of balancing. The spatial analysis of the locomotion paths between the POI is the second design activity the students engage in our approach (Fig. 6.1). It constitutes a filtering process, which selects a subset of suitable POG from the much larger set of POI.

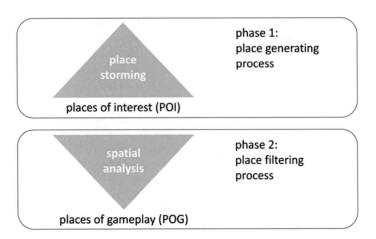

Fig. 6.1 Place storming and spatial analysis. Source: created by C. Schlieder

6.3   The Game Relocation Process

We adopt an extended version of the model of the geogame design process introduced by Schlieder and Kremer (2014) to describe the context of game relocation.
The model identifies six groups of design tasks and explains how the tasks interact
at the three levels of ludic design, narrative design, and performative design
(Fig. 6.2). It reflects the results of an analysis of a set of game design documents
created by the geogames team at the University of Bamberg.

The geogame design process begins with specifying the learning outcomes and
learning strategies that the game should support. Most importantly, the game has to
blend into the other activities in the curriculum. This requires the involvement of the
educator and some communication with the development team. In our own game
design projects, we sought early contact with the educators and asked them to view
the geogame as a computer-mediated field trip rather than as an extra-curricular
activity.

The two groups of design tasks, those relating to the game objectives and those
relating to the staging of the game, both refer to the situational context. They share
the fact that educators participate as co-designers, which is why the design process
model groups these tasks at the level of performative design. A sequential, non-
iterative design process such as the one indicated by the arrows in Fig. 6.2 starts and
ends with addressing issues of performative design. In practice, however, most
design teams iterate several times over the different groups of tasks. A walkthrough
of a simplified version of the process nevertheless provides a useful conceptual
framework for identifying how to streamline the design process for use in the
classroom.

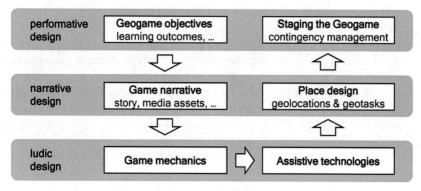

Source: created by C. Schlieder

Fig. 6.2 The geogame design process

The tasks of defining geogame objectives and staging the geogame at the *performative level* will be familiar to educators because similar issues are handled in connection with field trips, such as identifying goals and managing contingency plans, e.g. organizing student transportation or reacting to changing weather conditions. Just like during field trips, in our example students do not contribute to defining the learning outcomes nor game staging. Because educators are involved at this performative level, it constitutes a good starting point for simplifying student participation in the design process.

At the *level of narrative design*, we find two different groups of tasks. The task of designing the game narrative includes creating the game story line as well as the supporting media assets. In a typical educational geogame, the creative design team develops these elements in close interaction with a focus group of educators. Once the design phase finishes, the story line and the media assets stop to evolve and become an integral part of the geogame.

For the second group of narrative design tasks, however, the core design team will generally not be able to come up with a ready-to-play solution. These are the tasks concerned with *place design*, that is, with choosing the game's geographic content. Geogames tell spatial stories by selecting specific POGs and then associating place-related tasks to the POGs. Choosing places and tasks that have a good fit is critical for the gameplay component. For example, asking students to photograph a pine tree constitutes a trivial task in some geographic environments, but might be completely impossible in others. It follows that place design requires local geographic knowledge. As a design task, place design is accessible to students because they can use their everyday experiences with their spatial environment as a starting point. In our scenario, the students assumed the role of place designer by identifying the places of gameplay and creating the associated spatial tasks.

The *level of ludic design*, finally, groups the tasks that relate to the game mechanics and their implementation in mobile technology. *Game mechanics* is the term designers use to refer to all elements, which define the playing experience. Game mechanics include the rules defining the game, the modes of interaction with the player, and the game's incentive mechanisms. While we may expect most 12–16 year old students to have some experience with a variety of different game mechanics, this does not automatically make them good designers. Even students in a university-level game design program struggle with creating well-balanced game mechanics. For that reason, we did not ask the students to design the game mechanics, but let them build their game by re-using a proven mechanic. This approach has the additional advantage that neither the students nor the educators have to be involved with programming the software implementing the game mechanics.

Geogames are intrinsically related to places. This is why designers create, implement, and test them in a specific geographic environment. Once the geogames works there, the designers start to relocate it to other environments. In terms of the design process model, *geogame relocation* amounts to repetition of the place design process several times. Depending on the demands of the game mechanics, relocation may imply some changes in the game narrative, too. Typically, such changes do not concern the story line but they will affect place-related tasks and the media assets associated with them.

6.4   The CityPoker Game

The basic idea for simplifying the design process consists in letting the students, focus on the level of narrative design. Specifically, we treat the students as novice designers and focus on place design, providing them with solutions for the design tasks at the performative and the ludic levels.

We illustrate how to put the simplified design process into practice by using **CityPoker** as an example. This is a geogame designed by the first author, which the geogames team at the University of Bamberg has implemented for different mobile technologies. In **CityPoker** two teams of players compete to improve their poker hand by finding playing cards hidden in the geographic environment. Generally, the implementations of the game rely on a game server to synchronize the game flow, but there also exists a paper and pencil version of the game, which works with any form of direct communication between players (e.g. SMS, WhatsApp). The game narrative is as simple as any card game where the cards represent numerical values, not characters in a story. The details of the game mechanics are summarized in Box 6.1. The geogame borrows the concept of scoring hands from the variants of the traditional Poker card game. By its game mechanics, **CityPoker** is much closer to a non-betting Poker dice game than to Poker game variants such as Hold'em.

We concentrate on the task of place design and the role of the designers (students) in this task. We refer to the students as the "designers" keeping in mind that they are actually co-designers, since the game mechanics have been taken care of by the creators of the game. At the beginning of the game, the designers assign a hand of 5 playing cards from a deck of 20 cards to each team. The designers then hide

Box 6.1 CityPoker Game Mechanics
CityPoker is played with a deck of 20 playing cards consisting of 10, Jack, Queen, King, and Ace in the four suits ♠ ♥ ♦ ♣. In contrast to the traditional Poker card game, CityPoker is a full information game. Each team knows the hand of the opponent team. A map, which is updated during the game shows to both teams which caches contain which cards. A team may visit each of the five caches only once. Finding the caches involves a spatial search that depends on correctly solving a place-related task.

The game is played in real-time without turn taking. The teams move independently of each other taking the following decisions:

(1) identify the card that the team wants to obtain next in one of the five caches
(2) physically move to the cache's search region and solve the associated task
(3) if solved correctly, obtain information about the cache's exact location
(4) find the cache and trade in one of its cards

The game server updates the map to reflect the changes of cards in the cache and the new hand of the team. The game ends when both teams have visited all caches or after a maximum playing time, whichever occurs first. The team with the higher-ranking hand at the end of the game wins.

pairs of the ten remaining playing cards in five distinct places in the geographic environment. Place design amounts to identifying the places that may serve as caches for the playing cards. In designing a learning experience, the domain neutral narrative of the game has a significant advantage: it does not prompt competition for the learning outcomes. As an additional benefit, the domain neutral narrative needs no thematic adaptation. In other words, with **CityPoker**, the designers can concentrate on the task of place design.

A variety of learning tasks have been created to build on **CityPoker**. Schlieder and Kremer (2014), for instance, used the game to design a playful activity for geography classes about urban geography. The geogames team has gained a good understanding of the game mechanics and the factors affecting the balancing of the game flow, a considerable advantage to supporting the place design process. Furthermore, Kremer et al. (2013) found that the target age group of 12–16 year old students show considerable interest in this game.

In **CityPoker**, the caches are the POG. To qualify as POG, a place must come with an associated task that contributes to the learning objectives. In addition, the set of five places has to satisfy the requirements for a balanced game. The two features of the **CityPoker** game mechanics that have the most impact on place design are the full information scenario and the hierarchical spatial search. Both are game patterns, that is, reusable elements of the game mechanics, which may appear in

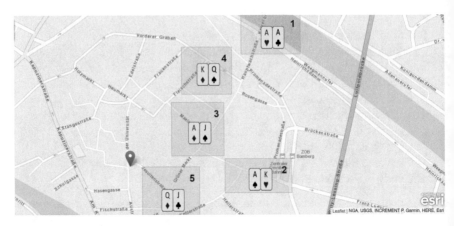

Fig. 6.3 The geogame design process. Source: generated by the CityPoker designer using ESRI cartography

other games, too (Björk et al. 2003; Sintoris 2015). The *full information pattern* is found in classic board games such as chess or checkers, where there is no hidden information. This pattern generally increases the combinatorial complexity of the game and causes players to reason strategically about how their game actions will affect the opponent's actions.

Figure 6.3 shows a typical distribution of cards as the **CityPoker** game client would show it. Note that the map displays rectangular search regions of different sizes, but not the exact cache location. The blue marker represents the start position for both teams. Consider a game, in which one team is dealt the hand K♠ Q♣ J♥ 10♣ 10♦. There are different strategic objectives this team might pursue. It could, for instance, try a full house (K♠ K♥ K♦ 10♣ 10♦) by trading in the required kings at caches 2 and 4. The opponent team, of course, will observe the team's actions and may prevent the strategy from succeeding by getting one of the kings first.

With respect to place design, the number of places that a strategy involves turns out to be the critical parameter. A game, in which one team just needs to visit a single nearby cache to obtain the best-ranking hand, is imbalanced. Both teams should have comparable chances of winning. A strategy that involves visiting all five caches risks being disrupted by the opponent team or the maximum playing time. Place design, thus, has to solve a spatial configuration problem: finding five places, for which the locomotion times of both team's optimal strategies are some-what balanced.

The second game pattern, *hierarchical spatial search*, relates the search regions to the caches. On the game map, the teams see only the search region, which contains the cache but not its exact location. To obtain the exact location, the team has to solve a quiz, whose answers are associated with different geographic locations. A typical place-related task is shown in Table 6.1. Exactly one of the four answers is correct. The associated location is that of the cache, the other locations being distractors. If the team does not know the answer, it has to search at all four locations,

Table 6.1 Place-related task

Visit one of the fruit and vegetable stalls located at the Green Market. What type of local produce do you find?		
1	Eggplants	*The cache is at the fountain on the Green Market.*
2	Asparagus	*The cache is behind the bench at Market Street No. 4.*
3	Oranges	*The cache is at the door of Market Street No. 18.*
4	Kiwi fruits	*The cache is in the flower box at High Street No. 24.*

which is going to slow it down considerably. Being able to exclude one or two answers reduces the number of places at which to search. Ideally, the team comes up with the single correct answer and can directly move to the cache. Note that any such quiz question is chosen specifically for the geographic environment in question. Answer 2 in Table 6.1, for instance, is true in the city of Bamberg at springtime, while it will be false in many other spatio-temporal contexts. Note also that the quiz has to be designed in such a way that there is exactly one correct answer. Table 6.1 shows the information as it would be presented in the paper-and-pencil version of **CityPoker**. The mobile game client directly displays the locations on the map without resorting to street addresses.

We call this game pattern *hierarchical spatial search* because it can be nested recursively. Instead of associating each answer with a location, it can be associated with another quiz question and so on. Again, the more answers are known in the hierarchical nesting of questions, the less time the players spend on spatial search. Although the authors of this chapter are not aware of other geogames making use of the hierarchical spatial search pattern, it is difficult to believe that no such game exists. The pattern is too simple and too useful as to not have been employed before. Having specified the **CityPoker** game we now move on to explain how to simplify place design, the central task at the narrative level of design (Fig. 6.2).

6.5 Narrative Design: Place Design Through "Place Storming"

The places of interest (POI) needed for an educational geogame are generally not the topographic landmarks or touristic sites that the student designers can retrieve easily from web-based repositories of point of interest data. A geogame links places to tasks. From the point of view of the gameplay, a place is interesting only if it supports one or more tasks that relate to a specific learning objective. Designing place-related tasks that are demanding and feasible at the same time poses some challenge. Place storming is a place generating method that helps the students to find suitable tasks by using the geographic environment as a creative stimulus and by providing a starting point for the associative process. The basic idea of place storming is *in-situ design*, that is, the requirement that the designer should be physically present in the POG.

Kristiansen (2009) has suggested an in-situ method for designing games called *site storming*. The first principle of his game design manifesto states: "site-specific games should be designed on-site". He argues that in this way, the designers gain a better understanding of the environmental constraints on the gameplay. An evaluation of the site storming method by Kristiansen et al. (2014) concludes that the method furthers creativity. Additionally, the participants found the design discussions which happened in the geographic environment particularly inspiring.

In adapting the site storming method for the purpose of place design, we depart in several ways from the original method. In contrast to site storming, our approach is based on group exploration and in-situ discussions in groups of students. In addition, place designers produce several designs—a task for each place—while the outcome of site storming consists in a single game mechanic. To distinguish it from site storming, we refer to our adapted method as place storming. We are aware of the fact that Anderson and McGonigal (2004) have used the same term outside game design for yet another in-situ brain storming method. However, introducing it with a different meaning in the context of geogame design should not cause confusion.

We consider a concrete learning scenario to illustrate the place storming method with an example. A geography class (student age between 14 and 16) works on the topic of sustainable food production. The students have spent some time in class researching geographic facts about food production and sustainability. A simple fact such as "seasonal local food has a small carbon imprint" serves as starting point for developing an in-game task. Similarly, the students could start from a geographic concept (e.g. carbon imprint) or a method (e.g. determining the carbon imprint). We call such a thematic anchor a *topic*.

Working in class, the students compile a list of topics to use in place design. The number of topics depends on the number of POI that the place generating process should provide, which in turn depends on the number of POG the game mechanics require. For instance, in **CityPoker** the game mechanics need five POGs. As a rule of thumb, the place generating process should identify at least twice as many, that is, ten POI in order to hand over a sufficient number of POI to the place filtering process, which chooses the POG. In other words, the students prepare at least ten topics about the field of sustainable food production.

The *topic inventory* created by the students constitutes the first resource for place storming. The topics are best seen as a to-do list to work on. In place storming, the students explore the environment and come up with a place-related task for each topic in the inventory. The outcome of place storming consists, thus, of a list of triple associations of a topic, a task, and a place. One such association for the sustainable food scenario is shown in Table 6.2. Note that in addition to the topic-place-task association, the table also lists potential results of the task, as they would be needed for the spatial search implemented by **CityPoker**.

In addition to the topic inventory, place storming uses an inventory of direct inputs to the creative process. The *task inventory* loosely corresponds to the game cards of site storming. It specifies a set of templates for tasks and gives hints on how

Table 6.2 Topic-task-place association

Topic	Seasonal local food has a smaller carbon imprint than out-of-season food because it requires less transportation and chilling.
Task	Visit a fruit and vegetable stall located at the Green Market that sells strawberries. Speak to the vendor and find out where and when the strawberries were harvested.
Place	The Green Market 49.89308°N, 10.88860°E
Answer	Harvested today at a place less than 50 km away
Distractors	Harvested 3 days ago at a place less than 50 km away Harvested today at a place between 50 and 200 km away Harvested 3 days ago at a place between 50 and 200 km away

to create variations of the tasks. A place-related task template can be as simple as "Ask a local person about X". In a variation of that template, the information asked for could be about an object shown on a photograph, which the players had to take in a prior phase of the game. We keep the task inventory deliberately small. It contains between 4 and 10 task templates (e.g. Ask someone about X, determine the geographic location of Y, collect data on Z).

Place storming needs only few preparatory steps. Box 6.2 describes the steps in form of a checklist. The general procedure of place in-situ brainstorming follow simple protocol. The two design teams A and B from step 5 of Box 6.2 explore the environment independently. Each team, however, moves as a group. Each item on the topic inventory is assigned to at least one member of the team. To make numbers work, some members may need to share the responsibility for a topic or, conversely, take care of more than one topic.

After the preparation phase, the teams start the spatial exploration. The student designers are instructed to carefully observe the environment and to pay special attention to places that relate to the topic(s) that have been assigned to them. When

Box 6.2 Checklist for the Preparation Phase of Place Storming

1. The educator has covered the general theme of the game (e.g. sustainable food production) with the students in class.
2. The students have compiled the inventory of topics, which they want to illustrate by place-related game tasks.
3. The educator has introduced the game mechanics (e.g. CityPoker) and explained how many POG it requires (e.g. 5 in the case of CityPoker).
4. The students have identified a local geographic region (e.g. the city's shopping district) where they want to set up their geogame.
5. The students have split into two design teams A and B. Team A is going to design the geogame that team B plays and vice versa.
6. The educator and the students have agreed on a date for the out-of-class activity and allocated a maximum time (e.g. 120 min) for place storming.

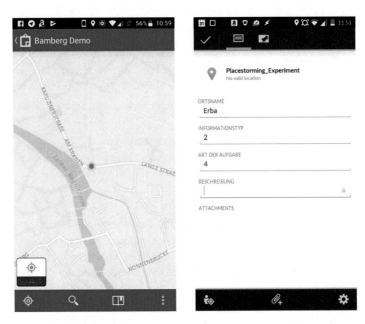

Fig. 6.4 Mobile documentation of topic-place-task associations. Source: generated by the Collector for ArcGIS using ESRI cartography

a student feels that the team passes a place of interest, he or she stops the group to find a suitable task. This is where the task inventory comes into play. The team member responsible for the topic selects two task templates from the inventory and discusses with the others what task variant would best match the topic. The team member then documents at least two task variants from two different templates and the group moves on. The exploration stops when the topic inventory is exhausted or the maximum time planned for place storming (see step 6 of Box 6.2) has elapsed.

Some of the parameters, of course, depend on the game mechanics. From our experience with place storming for **CityPoker**, we recommend for that game to have a topic inventory of ten items and a task inventory of just four items listing three variants for each. With 20 players in total, the teams are of size 10 and each team member takes care of exactly one topic. For an urban area with the dimensions of 500 m · 500 m, we recommend a maximum time limit for place storming of 120 min. In contrast to the place filtering process covered in the next section, place storming needs little technological assistance, if any. We found a tool such as the Collector for ArcGIS[1] helpful for documenting the topic-place-task associations (Fig. 6.4). Convenient features of a documentation tool are that it logs place coordinates automatically and stores the task descriptions on a server.

---

[1] http://www.esri.com/products/collector-for-arcgis.

We conclude the description of place storming with a remark on the cognitive demands of the tasks generated by student designers using that method. While most tasks will be as simple as the one shown in Table 6.2, designing the task is more complex even when a template is given since it involves metacognitive reasoning. Students will have to ask selves: Would I be able to solve the task at that place with the knowledge and skills acquired in the geography class? Note that the revised Bloom taxonomy of learning objectives considers metacognitive reasoning being a highly complex competence (Krathwohl 2002). Generally, the place storming tends to require higher order cognitive processes (analyzing, evaluating, creating) whereas the learning objectives of the play phase, that is, those of the tasks designed by the students, mostly relate to lower order skills (remembering, understanding, applying). Since each student engages in both, the designing and the playing phase, a broad range of learning objectives can be addressed.

6.6 Ludic Design: Game Flow Balancing Through the CityPoker Game Designer

The second phase of place design selects a suitable set of POG among the POI generated by the place storming method. This place filtering process differs in several respects from place storming: it is an in-class instead of an out-of-class activity and it is centered on spatial analysis instead of in-situ brainstorming. Most importantly, while place generation finds place-related tasks that match the learning objectives, place filtering tries to improve the balancing of the game flow.

Game research has studied the computational balancing of geogames because play testing the games in the geographic environment is extremely costly. Simple game mechanics like that of **GeoTicTacToe** can be analyzed by an exhaustive search of the game's state space (Schlieder et al. 2006). Probabilistic search is used for mechanics that are more complex. A computational analysis permits to identify place designs that guarantee fairness or prevent trivial winning strategies. Most of the results from game research require methods that are hardly accessible to students of the age range from our learning scenario (12–16 years). However, heuristics based on the findings from the computational analysis are often much simpler. In the remainder of this section we describe our adaptation of a method from computational game research, more specifically, an analysis based on the player model developed by our research group at the University of Bamberg (Kremer et al. 2013; Heinz and Schlieder 2015; Schlieder and Wullinger 2016).

Our simplified method addresses two fundamental issues of game flow balancing with a heuristic approach. The first issue is the maximum duration of the game and the second is the fair spatial distribution of the game resources. We inform the student designers about both issues and the heuristics used by the software tool we created (Box 6.3). The two heuristics constitute the design objectives for the place filtering process. Although we use analyses that are more sophisticated in our

Box 6.3 Instructions to Designers for the Place Filtering Process
As a geogame designer, you select the places of gameplay. You are responsible for ensuring it is possible to play the game with this selection of places. In particular, you should use spatial analysis to check the following two requirements.

1. Is the playing time acceptable? The average path from the starting point of the game through the places of gameplay and back should not take longer than 2/3 of the maximum playing time.
2. Is the game well balanced? The distance between two places of gameplay with strong playing cards should not be significantly smaller (<1/10) than the average distance between any two places of gameplay.

research, the heuristics give the student designers a good idea about the type of optimization that place design aims at.

Both heuristics are specific to **CityPoker**. However, the first is based on an observation that is valid for other geogames, too: the time players spend on locomotion dominates the time spent on other in-game tasks (Schlieder and Wullinger 2016). Designers generally want to minimize locomotion time. **CityPoker** is played with 5 POG, the places where the playing cards are hidden. Each team should have the opportunity to visit all 5 caches. Since the visit order depends on the team's plan as well as on the actions of the opponent team, we consider all possible 120 = 5! visiting orders. The upper diagram in Fig. 6.5 shows a typical distribution of locomotion time for the 120 visiting orders, which are ranked from the shortest to the longest. Note that the solution to the classical travelling salesperson problem is found at rank 1. The heuristic is based on the median of the locomotion times and requires that it should not exceed 2/3 of the planned maximum playing time leaving at least 1/3 of the time for the place-related tasks. If the maximum playing time is 60 min, the student designers have to choose the POG in such a way that the median of the locomotion times is 40 min or less.

The second heuristic relates to the spatial distribution of the playing cards. The cards considered strong and how this perception affects the game depends upon the strategies adopted by players. We used an agent-based simulation framework (Heinz and Schlieder 2015) to explore the effects of the strategies on different placements of the cards. Figure 6.5 shows a snapshot of a simulation run with the POG represented by pink polygons and the two player agents by green dots. The heuristics avoids putting strong playing cards in nearby places.

At this point, the students have completed the place generation phase and compiled an inventory of POI. This inventory lists at least ten places together with their topic-place-task associations and serves as a starting point for selecting the five POG needed for CityPoker. The **CityPoker Game Designer** or **CityPokerGD**, is a web-based application, which guides the student designers through the process of

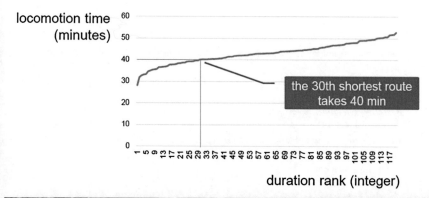

Fig. 6.5 Locomotion time and agent-based modeling

place filtering. We created the web application for use in the classroom, thus the students access the **CityPokerGD** using their preferred web browser on a notebook or tablet computer. The application guides the students through all phases of the place filtering process explaining the steps of the analysis and providing help functionality.

In **CityPokerGD**, we decomposed the process into a sequence of manageable subtasks that implement an iterative improvement process. The student designers start with selecting an initial configuration of POG from the inventory of POI. Depending on the results of the analysis of the initial configuration, the students discuss changing some of the places. If changes are necessary, the procedure restarts for the modified configuration of places and iterates until the spatial analysis returns a satisfying result. At the end of the place filtering process, when the students have identified the POG for the game, the **CityPokerGD** generates a descrip-

Table 6.3 Subtasks of place filtering

	Subtask definition	CityPoker Designer
1	*Geographic framing*: define the geographic environment of the game	Start point
2	*Place selection*: Choose a configuration of POG from the POI inventory	Cache placement
3	*Resource selection*: assign playing cards to the POG; define distractor locations	Card distribution Distractor placement
4	*Task selection*: assign place-related tasks for each of the POG	Cache field data
5	*Game field review*: reiterate steps 2–6 if the POG configuration needs improvement	Game field

Fig. 6.6 User interface of the CityPoker Game Designer

tion of the game. This is a configuration file for the game server or a PDF document for players who want to play the paper and pencil version.

Table 6.3 shows the subtasks supported and the titles given to them by the tool. The table lists the subtasks in the order that the tool guides the students through the process of place filtering. It is important to note that the decomposition into subtasks does not impose a strictly linear workflow. Students may address the subtasks 3–5 in a different order. They can also switch back and forth between the subtasks before having completed them. In practice, most students follow the suggested ordering of tasks in their first iteration, but move more freely in subsequent iterations.

The user interface of the **CityPokerGD** supports the subtasks with a specific input and feedback mode. Three elements define the interaction: an instruction panel, an interaction panel, and a feedback panel (Fig. 6.6). The instruction panel explains the interface elements that are visible as well as the actions the students have to take to complete the subtask. The students use the interaction panels to communicate decisions to the application and to input data. In the example of Fig. 6.6 it is an interactive map that serves as the interaction panel. Finally, the feedback panel informs designers of unresolved design problems. In many cases, it communicates the results of a spatial analysis that the **CityPokerGD** runs in the background. The feedback panel also informs about the state of completion of a subtask. Color-coded information boxes show at a glance which data has been provided completely (green) or partly (yellow), and which data is still missing (red).

Geographic framing: On opening the **CityPokerGD**, the users see a welcome screen, which explains the purpose of the application and provides a link to the rules of the **CityPoker** geogame. The students first define the geographic region of the game. They do so by either entering a city name, specifying a latitude and longitude, or clicking on a position of a web map. The first subtask also involves defining the start and end point of the game, a single place both competing teams move from at the beginning of the game and return to when the game is finished. Like all other design decisions, the position of the starting point can be revised at a later stage.

Place selection: In this subtask, the student specifies the configuration of places that he or she wants to use as POG for **CityPoker**. The places are input by locating them on an interactive map. As soon as a place is specified, the application performs several background computations. First, it computes the shortest pedestrian routes between the new place and the places that are already part of the configuration of POG. The application displays the routes on a map to help understanding the effects of the choice (see Fig. 6.6). With the visualization, the student can check the accessibility of the place for pedestrians, for instance. The second piece of information from the spatial analysis is the estimated playing time. Using the heuristics described at the beginning of this section, the application computes an upper and a lower bound, which is then displayed to the student designer. The application computes new routes and time bounds whenever the designer moves one of the POG to a new location. By giving immediate visual feedback, the **CityPokerGD** encourages playing around with different spatial configurations. This helps the students to anticipate which of the places from the POI inventory will actually improve the gameplay.

Resource selection: In **CityPoker**, the game resources are the playing cards hidden in caches located at the POG. The student designer assigns two playing cards to each of the POG. Once the cards are assigned, the **CityPokerGD** applies the heuristics for evaluating whether the game is balanced (Fig. 6.7). The application warns the student designer if extremely strong pair of cards are located too close to each other. In addition, distractors need to be defined. Although distractors are geographic locations, they are more similar to resources than places, and they are required for the hierarchical spatial search pattern that our game uses. The distractors are always located in the vicinity of a POG, and their exact position does not have a big effect the gameplay as far as the heuristic evaluation from Box 6.3 is concerned. With respect to the distractors, the application checks that the student designer has defined them all.

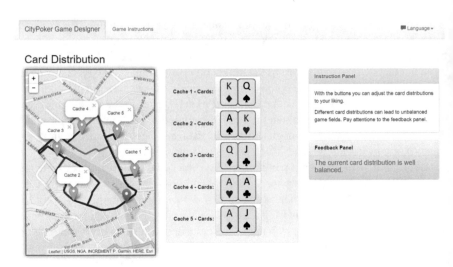

Fig. 6.7 Card distribution

Task selection: This subtask can be seen as mere data input step since it does not involve decisions that affect the balancing of the game. The student just copies the task description from the POI inventory into a **CityPokerGD** and assigns possible results of the task to the POG and the distractors. A design decision is taken only in cases where the POI inventory contains alternative tasks associated to a place. Even in such a case, the decision will affect the game play but not the spatio-temporal balancing.

Game field review: In this last step, the student designing the game reviews all of the game data. He or she decides whether to continue improving the place design or to finalize the geogame.

Although the students could work independently with the **CityPokerGD**, working in groups stimulates discussions and generally produces better results. Based on our experience with the design process, we recommend that each of the two design teams splits up in groups of 3–4 students. The games produced by the groups of one team will generally differ in the choice of the POG as well as in the task associated with the POG. In a final team meeting, the members decide which of the results becomes the team's geogame that is going to be played by the other team.

6.7 Discussion and Outlook

We have covered quite a lot of ground in this chapter. For the final discussion, let us return to our starting point: designing a geogame in the classroom. While teachers have used geogames in a variety of learning contexts in secondary education, they generally avoid letting the students themselves design the game because of the alleged complexity of the task. In this chapter, we have identified an important part of the geogame design process, namely game relocation and, more specifically, place design, which is at the same time sufficiently challenging and not too complex to be of interest in an educational context. We provided the methodological and technical means for addressing place design in the classroom.

Now that we have detailed our approach, it is apparent how the three geogame design principles to enhance student learning stated at the beginning of the chapter are implemented. The first principle asked for starting with a rule set known to produce a well-balanced game. We did this by applying place design to the **CityPoker** geogame. The second principle consisted in letting the students explore the geographic environment of the game. With the place storming method, we provided a solid basis for this phase of the design process. Finally, we showed how to optimize the game flow as stated in the third principle. We formulated heuristics for spatial analysis and we created a software tool, the **CityPokerGD**, to implement the heuristics.

All of these elements together provide a blueprint for a class on geogame design. This has been our primary goal. However, our analysis also contains material that is neither specific to a particular geogame nor to geogame design education. We are convinced that the design process model we presented will prove useful in other application scenarios, too. Most importantly, the model helps to understand the importance of place design in game relocation. Game relocation has evolved into a topic of active research, and is also addressed by other chapters of this book. However, as described in this chapter, place design has received much less attention. Place design is especially challenging because designers cannot look exclusively at the spatial constraints of the problem. In moving a successful game from one geographic environment to the other, there are other aspects to consider, most notably, social and cultural constraints. This relates place design to the recent research activities on place-based GIS, which also aim at capturing the social meaning of spatial entities.

Acknowledgments This research has been funded in part by ESRI Inc. within the Geogames and playful Geodesign project.

References

Adams E, Dormans J (2012) Game mechanics. Advanced game design. New Riders, Berkeley, CA

Althoff T, White RW, Horvitz E (2016) Influence of Pokemon go on physical activity: study and implications. J Med Internet Res 18(12):e315. Preprint arXiv:1610.02085

Anderson K, McGonigal J (2004) Place storming: performing new technologies in context. In: Proc. Nordic conf. on human-computer interaction. ACM Press, New York, pp 85–88

Björk S, Lundgren S, Holopainen J (2003) Game design patterns. In: Raessens J, Copier M, Goldstein J, Mäyrä F (eds) DiGRA-03: conference on digital game research. Digital Game Research Association, Utrecht, NL

Celtek E (2010) Mobile advergames in tourism marketing. J Vacat Market 6(4):267–281

De Gloria A, Bellotti F, Berta R (2012) Building a comprehensive R&D community on serious games. In: VS-GAMES-12, proceedings of international conference on games and virtual worlds for serious applications, pp 1–3

de Souza e Silva A (ed) (2009) Digital cityscapes. Merging digital and urban playspaces. Lang, New York, NY

Heinz T, Schlieder C (2015) An agent-based simulation framework for location-based games. AGIILE-16 proceedings of international conference on geographic information systems. https:// agile-online.org/conference_paper/cds/agile_2015/shortpapers/114/114_Paper_in_PDF.pdf

Hodson H (2012) Google's ingress game is a gold mine for augmented reality. New Scientist 216:19

Klopfer E (2008) Augmented learning—research and design of mobile educational games. MIT Press, Cambridge, MA

Krathwohl D (2002) A Revision of Bloom's Taxonomy, Theory into Practice 41(4):212–264

Kremer D, Schlieder C, Feulner B, Ohl U (2013) Spatial choices in an educational geogame. In: Proceedings of VS-Games 2013 (5th international conference on games and virtual worlds for serious applications)

Kristiansen E (2009) Computer games for the real world: designing a design method for site-specific computer games. PhD thesis. Roskilde University. pervasivegames.files.wordpress. com/2010/05/ekfinal2.pdf

Kristiansen E, Pries-Heje J, Baskerville RL (2014) Designing scientific creativity. In: Proceedings of Scandinavian conference on information systems. Springer, Berlin, pp 44–57

Lude A, Schaal S, Bullinger M, Bleck S (2013) Mobiles, ortsbezogenes Lernen in der Umweltbildung und Bildung für nachhaltige Entwicklung (in German: Mobile, locaton-based learning in environmental education). Schneider Verlag, Baltmannsweiler

Montola M, Stenros J, Waern A (eds) (2009) Pervasive games. Theory and design. Morgan Kaufmann Publishers, Burlington, MA

Schaal S, Grübmeyer S, Matt M (2012) Outdoors and online-inquiry with mobile devices in pre-service science teacher education. World J Educ Technol 4(2):113–125

Schlieder C, Kiefer P, Matyas S (2006) Geogames: designing location-based games from classic board games. Intell Syst IEEE 21(5):40–46

Schlieder C, Kremer D (2014) Geogames: Schüler entwickeln ein ortsbezogenes Spiel (in German; Geogames: students develop a location-based game). Praxis Geogr (7/8):31–35

Schlieder C, Wullinger P (2016) Reducing locomotion overhead in educational geogames. In: Gartner G, Huang H (eds) LBS-16, proceedings of international conference on location based services. ACM, New York, pp 228–232

Sintoris C (2015) Extracting game design patterns from game design workshops. Int J Intell Eng Inf 3(2-3):166–185

Sinton D, Lund J (eds) (2007) Understanding place understanding place: GIS and mapping across the curriculum. ESRI Press, Redlands, CA

von Borries F, Walz S, Böttger M (eds) (2007) Space, time, play: computer, games, architecture, and urbanism. Birkhäuser, Zürich

Chapter 7
(Re-)Localization of Location-Based Games

Simon Scheider and Peter Kiefer

7.1 Introduction

Location-based games (LBGs) involve movement in and large-scale interaction with *environmental space* (Nicklas et al. 2001; Schlieder et al. 2006), which is the space larger than the body which cannot be comprehended without considerable locomotion (Montello 1993). They form an important subclass of *mixed reality games*, i.e., computer games played in a physical environment which add novel dimensions to the game experience, including seamless *immersion* of players, new kinds of *social interaction* with other players, as well as *physical interaction* with the environment (Hinske et al. 2007). The main advantage of such games is that physical and social experiences are most authentic in a concrete physical or social environment, while the virtual layer of mixed reality adds unprecedented forms of imagination to these environments. In *pervasive games*, the virtual, social, and physical environments are interconnected based on weaving computing power and sensors into the environmental fabric, and based on the fact that players constantly carry mobile devices (Hinske et al. 2007; Benford et al. 2005; Walther 2005). We regard LBGs as a particular subclass of *geogames*, i.e., games played in geographic space (Schlieder et al. 2006). The latter, however, include also online games that make use of geographical information without any physical interaction of players (Ahlqvist et al. 2012).

S. Scheider
Institute of Cartography and Geoinformation, ETH Zurich, Zurich, Switzerland

Department of Human Geography and Spatial Planning, University Utrecht, Utrecht, Netherlands

P. Kiefer (✉)
Institute of Cartography and Geoinformation, ETH Zurich, Zurich, Switzerland
e-mail: pekiefer@ethz.ch

© Springer International Publishing Switzerland 2018
O. Ahlqvist, C. Schlieder (eds.), *Geogames and Geoplay*, Advances in
Geographic Information Science, https://doi.org/10.1007/978-3-319-22774-0_7

While LBGs have been around for some time (Nicklas et al. 2001), only few of them have succeeded in attracting a larger number of players. One reason is the difficulty of embedding game concepts in an environment. In order to reach players from different places and in order to allow for flexibility in taking gaming opportunities, LBG concepts need to be easily re-localized in a way which preserves the particular attractiveness of a game. Furthermore, turning successful virtual reality games played on a computer, or massive multiplayer online games (Ahlqvist et al. 2012), into a LBG requires localization, i.e., the suitable embedding of virtual game concepts into a physical environment. All these tasks still pose considerable conceptual and computational challenges, even though some effort has been made to tackle them (Schlieder et al. 2006; Kiefer et al. 2007; Hajarnis et al. 2011; Schlieder 2014). Furthermore, it recently has become popular to use game concepts in non-game contexts for persuasive computing (gamification) (Deterding et al. 2011; Scheider et al. 2015). Here, too, the successful embedding of elements of games and play into an environment constitutes a considerable challenge for designers (Hassenzahl and Laschke 2015).

In general, we can distinguish three research challenges on the way towards really flexible location-based gaming:

1. How can (arbitrary) games be localized?
   (*Game* → *Game* + *Env*)
2. How can location-based games be re-localized?
   (*Game* + *Env*$_1$ → *Game* + *Env*$_2$)
3. How can environments be gamified?
   (*Env* → *Game* + *Env*)

In order to provide answers to these questions, and to facilitate corresponding game localization technology, it is necessary to develop *computational quality criteria for the embedding of games in an environment*. While this problem has partly been recognized in the literature (Schlieder 2014), a systematic derivation of criteria which take into account a game's ludic dimension, the game narrative, as well as the activity-based embedding into an environment, is still missing.

In this chapter, we discuss the problem of game localization in the light of recent game literature and environmental and psychological models of space (Sect. 7.2). Based on this, we propose a layered (3-tier) model of game localization (Sect. 7.3) which provides a way of addressing all three questions introduced above. We use this model to suggest some novel quality criteria (Sect. 7.4) for games which particularly reflect their environmental embedding and are based on state transition graphs. We illustrate our criteria with a hypothetical conquer game that has a very simple state transition graph (Sect. 7.5), and discuss its application to an existing LBG (Sect. 7.6). We conclude the chapter in Sect. 7.7 by discussing in how far our method provides answers to the research questions posed above, and what still needs to be done.

7.2 Location-Based Game Concepts and Related Work

Games consist of different conceptual elements which deploy a game in environmental space. These elements determine its quality, and thus need to be taken into account in game localization.

7.2.1 Games and Play

The element of *play* refers to a kind of embodied activity which is shared and involves social roles, and which is deeply rooted in human biology (Stenros 2015). Play ranges from foundational forms of hiding and chasing to sophisticated forms of role play in a theater. The main characteristic of play is that involved objects and agents can play roles different from what they are supposed to be (outside play), and that the rules which guide play are not explicit, fixed and shared (Stenros 2015), i.e., they are not institutionalized facts (Searle 1995). Games, in contrast, can be seen as an institutionalized (codified) form of play (Stenros 2015), where (collective) intentionality presupposes that players stick to certain rules and follow pre-defined goals. Play accounts to a large extent for the experience of *immersion* and *flow* in a game (Hinske et al. 2007), where large parts are probably not made explicit or happen on a subconscious level. The explicit restrictions and rules that come with a game sometimes can even destroy a *playful experience*, partly because breaking and redefining the rules is an intrinsic part of play (Stenros 2015). Still, a game retains an essential part of the play experience in the form of activities and roles which allow players to connect a game to meaningful places, scripts and narratives in the environment. We therefore hold that play is an intrinsic part of localizing games.

7.2.2 Scripts and Narratives in Games

In classic game research, there is a debate between ludologists, who investigate games in terms of game mechanics, referring primarily to their rules and winning strategies and sometimes denying the relevance of narratives in games, and narratologists, who see games primarily as a form of interactive story telling (Jenkins 2004). While games admittedly work in a different way than *plots* in cinema or fiction, in the sense that the story is not told linearly and is not (entirely) in the hands of the game designer, narratives do play an essential role in game localization (Paay et al. 2008). The reason is that in LBGs, players often understand the environment in terms of a narrative, and thereby project the game onto the environment. This narrative has *a non-linear spatial form* (Jenkins 2004), based on roles for things distributed in space that can be accessed by a player. In this way, games can evoke collectively known stories, such as pirate stories in a Disney amusement park.

Furthermore, players can push a story forward by movement in space (e.g., when a story unfolds through space, as in Bichard et al. (2006)), by revealing background stories (e.g., the murder in a classical detective game), or by constructing emergent stories on their own as in a game like **The Sims** (cf. Jenkins 2004), and through this they are able to break the linear narrative. For example, in backseat games (Bichard et al. 2006), where players move through an environment in the backseat of a car listening to a detective story that plays in their surroundings, the background story can be actively pushed forward at certain locations in that environment. Even though some games may not involve an elaborate linear plot, we suggest that game localization is always a matter of the design of a *spatial narrative* (Jenkins 2004), where either some roles or some points in the story are fixed to locations, objects or activities in environmental space. In some cases, stories may be reduced to a minimal form, such as a *script* (a stereotyped sequence of events) or a *frame* (a stereotyped situation) in cognitive linguistics (Petruck 1996). In these cases, roles may be almost unrecognizable and remain manifest only in the names of figures, such as the queen game piece in chess.

7.2.3 Places and the Meaningful Environment in Games

LBGs need to control the space in which players act (Lemos 2011). This was first discovered by researchers on pervasive gaming: Benford et al. (2003, 2005) raised the problems of uncertainty, spatial configuration, and temporal orchestration of a pervasive game, which are caused by its embedding in space. Walther (2005) distinguished tangibility space, information space, and accessibility space, where the first is the space of possible interactions with a physical environment, the second is a digital game representation of the first, and the third interfaces the former two. Several authors (de Souza e Silva 2008; Montola 2005) argued that LBGs are performed simultaneously on different virtual, social and physical spaces which extent the "magic circle" of a game to encompass "serious" social life activities, and thus extend cyberspace to *geographic places* and *objects* (Lemos 2011). Reid argued that all LBGs have a degree of place-related embedding, which corresponds to the extent to which their narratives specifically relate to existing places instead of only loosely overlapping space (Reid 2008).

In the age of digital information, space is often reduced to GPS coordinates. Place, in contrast, appears to be a more involved category of Geography (Cresswell 2013), which is closely related to daily activities (Seamon 1979), routine habits as well as narratives (Tuan 1977, 1991). Places shape possible actions (affordances) (Scheider and Janowicz 2014) by their spatial layout, by the people who live there, as well as by convention. In this, they are comparable to Gibson's meaningful environment (Scheider and Janowicz 2010; Gibson 2013), which is a way to regard an environment in terms of what it affords to animals or humans. For these reasons, mobile technology needs to take existing places into account (Dourish 2006). Designing games such that player interactions closely correspond to affordances of those places in which they are played increases a player's immersion and feeling of authenticity, and thus, gives meaning to ludic activities.

The latter seems, however, an ongoing challenge for game designers. Most pervasive games to date are rather "spatial" than "platial": they largely consist of chases and hunts (Lemos 2011), and the interaction between space and cyberspace is reduced to tracking unrestricted movement or to arbitrary space division. For example, a game like **Parallel Kingdom**[1] arbitrarily divides geographic space into territory claims, without taking into account the structure of existing places. Another example is **Zombies Run!**,[2] in which joggers can escape Zombies by running in any direction. In Google's successful contemporary LBG **Ingress**,[3] urban landmarks form "portals" which need to be "hacked" and "linked" to generate "control fields", i.e., spatial triangles under the control of a group of players. However, the choice of landmarks and triangles is arbitrary, and there is no dependency between player actions and geographic places, in particular since actions remain largely virtual.[4]

7.3   A Layered Model of Game Localization

While existing models of game localization mostly focus on a game's codified rules or technical infrastructure for ubiquitous computing, playing a game is always a fabric of roles, concepts and actions on different layers of conceptual abstraction embedded into an environment. Large parts of these layers are often not made explicit or represented in a computer. In fact, one may consider LBGs as primary examples for the mingling of digital and analog computation (MacLennan 2009), in which the human environment takes over important roles in activities not necessarily represented in a digital form.

7.3.1   Three Conceptual Game Layers

Following the suggestion of Schlieder (2014), we distinguish the *ludic, narrative* and *environmental* layer (see Table 7.1). The ordering of layers in Table 7.1 is important here, since lower ones are assumed to *deploy* or *implement* concepts of the upper layers. For example, a building in the environment may play the role of a castle on the narrative layer and be simply a place for resources on the ludic layer.

---

[1] http://www.parallelkingdom.com.

[2] https://zombiesrungame.com/.

[3] https://www.ingress.com/.

[4] This problem has recently led to serious ethical complaints of the German public. Ingress players had erected portals inside the former concentration camp Sachsenhausen in Berlin, cf. http://www. zeit.de/zeit-magazin/leben/2015-07/ingress-smartphone-spiel-google-niantic-labs-kz-gedenkstaette.

Table 7.1 The three conceptual layers of a LBG

Abstraction level		Actions	Constraints	Quality criteria
0	Ludic	Game actions (e.g. re-allocation)	Game rules and mechanics	Game balancing
1	Narrative	Play actions (e.g. conquer)	Scripts and story	Authenticity
2	Environmental (perception or simulation)	Environmental actions (e.g. movement)	Affordances	Playability, breakability

The action of walking somewhere on the environmental layer may correspond to an invasion on the narrative layer and an ownership change of a place on the ludic layer. To account for the dependency between layers, a *mapping* between layers becomes necessary, which is discussed later in this chapter (see Sect. 7.4.1).

We furthermore assume that layers on higher levels are not reducible to lower levels because each layer adds specific *constraints* to the game actions, which are not necessarily present on other layers. For example, the ludic layer adds constraints by codified game rules (such as, whether a player is allowed to go to a place), the narrative layer adds constraints concerning scripts and roles in a corresponding narrative (such as, kings need to travel in carriages), while the environmental layer adds constraints concerning what can be done (affordances) in an environment (e.g., reaching a place in a certain time). Each layer thus adds *quality criteria* for games associated with its constraints and concepts (see last column of Table 7.1).

In the following sub-sections, we suggest a *state transition model* for the layers in this hierarchy. This model is the basis for the localization criteria described in Sect. 7.4.

7.3.2   Ontology of Game States

During game play, all layers are in a *game state*. States are described by sets of facts present on a given layer. *Actions and other processes* can change this state from one to the next. Figure 7.1 gives an overview of a simple game state ontology expressed in OWL.[5] We suggest this game state ontology as a *pattern* (Gangemi and Presutti 2010), i.e., a minimal ontology required to describe a LBG on the ludic layer. Note that more specific classes can be introduced for a specific game, and that narrative and environmental layer will extend this ontology.

Among the *classes* of this OWL pattern, we have *Agent* which denotes the set of things that can act intentionally, and *Player* as a subclass of Agent which encompasses all agents that participate in a game. *Object* denotes the set of things that are neither agents, places nor locations. *Place* and *Location* localize games and need to be distinguished in order to cope with both discrete, cognitively meaningful space

[5] http://www.w3.org/TR/owl-features/. This is the "Web Ontology Language", a W3C recommendation for describing Web information with ontologies.

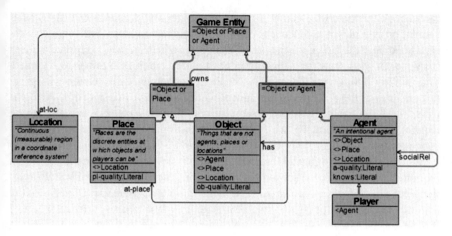

Fig. 7.1 An OWL ontology of the classes and properties used to describe the state of a LBG

(such as cities and market places), as well as continuously measurable space (e.g., in terms of GPS coordinates in a spatial reference system).

Among the *properties* (denoting binary relations), we distinguish *owns*, which denotes an agent's ownership of some object or place, *has*, which denotes that an agent carries some object, *at-place*, which denotes that agents or objects are located at some place, and *at-loc*, which denotes that agents, objects or places are located in some coordinate region (which may also be a single point). The union of the latter two properties is simply called *at*. The distinction of ownership and possession can be important but may be irrelevant for a particular game. We also introduce properties which assign attribute values (*qualities*) to objects, places, and players. These can be used to model all kinds of unary states, including also states denoting events. For example, the fact that an agent knows something relevant for the game can be modeled as a quality[6]. Furthermore, we distinguish a single property *socialrel* among players which denotes possibly diverse social relations between them, such as that they belong to the same group, are at war, or that one player is superior to another player.

7.3.3 Game Processes as State Transitions

Game processes are modelled as state transitions, i.e., operations which trigger *changes of the sets* denoted by the properties and classes of the state ontology on the respective layer. Note that game states may change with and without player actions involved. In principle, one can therefore distinguish two kinds of processes which

---

[6] Knowledge is modeled here simply as a particular state, without taking into account any more sophisticated (modal) logic.

can change the state of a game: a *game simulation* and a *game sensing*. A game simulation is a player-independent computer simulated process. For example, at a certain point in the game, a sequence of changes can be triggered to enforce a linear storyline, or there may be a random generator that enforces changes to a game's state, similar to throwing a dice. Game sensors, in contrast, detect changes in qualities or states of the perceived environment which cannot be influenced by the computer, such as whether some object moves around, detected by positioning technology. In the following, we do not further distinguish between these two kinds of processes.

We use another ontological hierarchy for describing kinds of state transitions. In the following notation, the subsumption operator $\sqsubseteq$ denotes the "*subClassOf*" relation between *state transition classes*. Note that *individual* state transitions form the *edges* of a state transition graph, whereas state transition classes form the *labels* of these edges (see Sect. 7.4.2). Actions, whether performed by players or not, are the most important kinds of state transitions in games:

$$action \sqsubseteq stateTransition$$

When an agent decides to perform an action, this—in essence—changes at least one of the sets which describe a game's state. We can specify an action type therefore by the sets it is supposed to modify, using the symbol $::\rightarrow$[7]:

$$(changeowner ::\rightarrow owns) \sqsubseteq action$$

$$(take ::\rightarrow has, \ at) \sqsubseteq action$$

$$(put ::\rightarrow has, \ at) \sqsubseteq action$$

$$(move ::\rightarrow at) \sqsubseteq action$$

$$(changesocial ::\rightarrow socialrel) \sqsubseteq action$$

$$(learn ::\rightarrow knows) \sqsubseteq action$$

We distinguish ownership change from reallocation (take, put), since ownership change is possible without any location change. Movement, in contrast to reallocation, denotes only movement of players. Learning something means that some agent gets to know something.

---

[7]This is an informal notation, which illustrates the usage. A formal notation would make use of corresponding transition rules, see below.

7.3.4   Ludic Layer

On the ludic layer, a game has a set of codified (shared and institutionalized) rules, i.e., the *rules of a game*, which constrain player actions that modify a game's state. Ludic game states are modelled therefore in the simplest possible form sufficient to describe such ludic constraints.

The rules of a game are specific to a game, and thus cannot be specified in general. Codifying a game's ludic rules can be done in terms of *inference rules*, denoted by $\Rightarrow$, specifying the conditions for actions (in the rule body) as well as their outcomes (in the rule head). For example, the action type *take* can be defined as follows:

$$take : Object(x),\ Agent(a),\ at(x, p),\ at(a, p), \Re has(a, x)$$

$$\Rightarrow has(a, x),\ \neg at(x, p)$$

We assume that all ludic player actions are made explicit, since otherwise, it is not possible to compute a *state transition graph* on the ludic layer, i.e., a graph which explores action possibilities in an exhaustive form.

Besides the rules of a game, the ludic layer also qualifies particular states of a game, namely *starts* and *goals*, based on corresponding *start and win conditions*. For each player, game states are evaluated according to a win condition. For example, the goals of some games are based on a score of ownership, such as Monopoly, while others are based on a geometric state condition, such as checkmate in chess.

7.3.5   Game Narrative

On the narrative layer, *classes* and *properties* are added to the ontology which embed a game state into a certain narrative or script. For example, a fantasy game may add the following subclasses and properties to the game state ontology:

$$Wizard \sqsubseteq Agent$$

$$Dwarf \sqsubseteq Agent$$

$$Witch \sqsubseteq Agent$$

$$superior \sqsubseteq socialrel$$

This specifies that, in this example, three different kinds of agents participate in the game, and that there is a particular type of social relation *superior*, which denotes whether somebody was superior in a fight.

Also new state transitions (including player actions) can be added which are specific to this narrative, e.g.

$$walk \sqsubseteq move$$

$$ask \sqsubseteq learn$$

$$(attack ::\rightarrow superior) \sqsubseteq changesocial$$

$$conquer \sqsubseteq changeowner$$

Similar to the ludic rules above, on the narrative layer these actions may be further constrained. For example, in our fantasy play, players may be able to ask somebody only if the other Agent is spatially present. Furthermore, one may only be able to conquer something from somebody if the superior relation holds, e.g., as a result of an attack action. Note that narrative constraints may not be necessary for playing the game on the ludic layer, but still add a sense of authenticity and can account for large parts of the play aspect of a game. In this chapter, we treat narrative constraints as (non-codified) rules in a similar way as ludic constraints.[8]

In a classical computer game, almost all game actions and states on higher layers map into virtual actions and states in an environmental simulation of the game, including virtual layouts visible on the computer screen. The only kind of physical action involved may be joystick manipulation and screen interaction. In a LBG, many actions translate into physical movement or manipulation of objects in the perceived human environment. The degree to which this is the case determines the degree to which a game is a *location-based*, and thus determines its spatial scope.

The constraints on the layer of environmental perception are usually not made explicit on a computer. Actually, they are given by *environmental affordances*, and thus are implicit in the relation between objects and environment (Gibson 2013).

7.3.6   Environmental Perception and Simulation

The environmental layer is the level of direct *player interaction* with a game, i.e., of interactions between physical and virtual entities through appropriate sensory interfaces. The *perceived environment grounds* the upper layers (c.f. Scheider 2012), i.e., it serves as spatial anchor for the game abstraction hierarchy. It contains objects and layouts as well as corresponding affordances and actions, as proposed by Gibson (2013).[9] *Environmental simulation*, in addition, denotes the computer simulation of the environment in a game. It can contain exactly the same kinds of things as the

---

[8] Whether this strategy is always applicable seems an open question of research: can narratives always be formalized in terms of rules?

[9] Environmental perception, as a matter of fact, can be considered a kind of simulation performed by our brains (cf. Hawkins and Blakeslee 2007; Scheider 2012).

perceived environment or further ones, such as ghosts or monsters. Just as the perceived environment, it displays game affordances and thus serves as a gaming interface for player actions. If the two environments are blended over each other, they constitute an augmented reality.

7.4 Game Localization Criteria

In essence, game localization criteria are a function of the *particular embedding* of a layer into lower layers, taking into account the *constraints*, which exist on each layer. In the following, we discuss localization as embedding and the preservation of consistency under state transitions, define a number of novel criteria based on embeddings, and discuss how these criteria may be measured and computed.

7.4.1 Game Localization as Embedding

Localizing a game means to establish mappings between the narrative and the environmental layer, as well as between the narrative and the ludic layer—both, for kinds of state transitions, as well as for those entities, classes and properties (represented as unary and binary relations, respectively) that describe the state of a game (see Fig. 7.2).

Mappings need to establish identity between layers in a way that still allows for layer-specific modifications of facts. For example, the fact that a knight is located at some forest on the narrative layer may translate into the fact that some player is located at some park on the environmental layer, or the action class horse riding may be translated as tram riding. Furthermore, we require that every fact describing the state of a game on higher layers and every state transition class is translated into lower layers, and thus into the environmental layer. That is, the mapping needs to be *total*. This is because a game's state needs to be fully controlled bottom-up by environmental processes, regardless of whether they are triggered by player actions, non-player processes or simulations. We leave open whether mappings are established ad-hoc, i.e., during the playing of a game, or a-priori.

The main purpose of the mapping is to pinpoint those entities in an environment (or in a narrative) which are supposed to play a role in the game. As depicted in Fig. 7.2, we can identify game-relevant things on each layer in terms of the respective *images* of the mappings. There may be other things on each layer that do not play a role in a game (e.g., smoking as an action on the environmental layer). Furthermore, we do not require mappings to be *injective* (one-to-one), because there may be objects of the environment playing several roles in the game, and because there may be several ludic/narrative processes that map to a single process in the environment. For example, both swimming and riding in the narrative might be mapped to walking in the environment, whereas a ruin in the environment could be

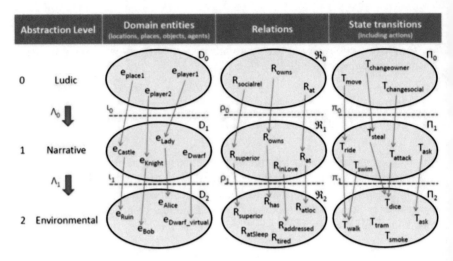

Fig. 7.2 The principle of game localization as a mapping of the three sets of game elements: domain entities (*D*), relations (R), and state transitions (Π) between the three layers

used for two different castles on the narrative layer. We thus propose to map game elements *top-down*, i.e., from ludic to narrative and from narrative to the environmental layer.

In summary, we propose that game localization consists of a mapping (refer also to Fig. 7.2):

$$\Lambda = \{\Lambda_0, \Lambda_1\}, \text{ where}$$

$$\Lambda_i = \{\iota_i, \rho_i, \pi_i\}, \text{ with}$$

$$\iota_i : D_i \mapsto D_{i+1} \text{ and } \rho_i : \mathfrak{R}_i \mapsto \mathfrak{R}_{i+1} (\text{total})$$

$$\pi_i : \Pi_i \mapsto \Pi_{i+1} (\text{total})$$

Here, the index $i \in \{0, 1\}$, with 0 = ludic, 1 = narrative, 2 = environmental layer, which means that the mapping stops precisely when mapping from the narrative into the environmental layer. The domain and range symbols of these mappings have the following meaning

$$D_i := \text{set of entities (domain) on layer } i$$

$$\mathfrak{R}_i = \{R_{i1}, \dots, R_{in}\} := \text{set of relations on layer } i$$

$$\Pi_i = \{T_{i1}, \dots, T_{im}\} := \text{set of state transition classes on layer } i$$

with i ranging this time over all three layers, and n and m denoting the sizes of corresponding sets.

7.4.2 Consistency Preservation of Game States Under a Given Localization

A mapping of a certain game state into lower layers can be *consistent* or not. We define a consistent mapping as one that preserves states of affairs between layers. This is also called a *homomorphism*. If the mapping of game states is homomorphic, then it is the case that:

$$R_{ij}(a,\ldots,z) = \rho_i(R_{ij})(\iota_i(a),\ldots,\iota_i(z)) \text{ for all } R_{ij} \in \mathfrak{R}_i \text{ (homomorphism)},$$

where a to z denote individual things and R_{ij} the j - th relation on layer i, and ρ_i, ι_i denote mappings as defined above. This *propagates states of affairs* upwards from the environmental layer to higher layers. For example, if ownership change on the narrative layer is translated as taking some object on the environmental layer, then whenever I have taken an object, a homomorphic mapping would cause me to own that object (Fig. 7.2).

Since we do not require a localization to be homomorphic, and since state transitions are bounded by *independent constraints* on each layer, a game state can become inconsistent. Figure 7.3 illustrates two inconsistent states in a state transition graph. The only consistent state is the start (state 1), while the two depicted follower states (states 2 and 2′) are inconsistent: if we map the relation *owns* on the narrative layer to the spatial relation *at* on the environmental layer (Fig. 7.3), and if it was excluded by narrative rules that two people can own a castle at the same time, then every time two people move to the ruin which denotes that castle (such as Peter and Bob in Fig. 7.3), the game state becomes inconsistent (state 2). More precisely, players can move on the environmental layer in a way which enforces a state transition on the narrative layer that *breaks the rules*. We call this possibility of generating an inconsistency in a game *breakability*. And vice versa, suppose that based on narrative constraints, we compute possible moves of a player, and that one of these possible moves leads a player straight across a wall (such as Bob in Fig. 7.3). Since this move is excluded by affordances on the environmental layer, it leads to a state inconsistent with environmental constraints (state 2′), and thus to a state which is not playable. The non-playable subset of the narrative graph therefore consists of edge 1 and state 2′, and the breakable subset of the environmental graph consists of edge 2 and state 2.

It is important to understand that these possibilities of independently pushing a game into inconsistent states on different game layers under a given localization is exactly what causes *state transition graphs* to become *incompatible* between layers, and thus games to become either increasingly *non-playable* or *breakable*. The *non-playable subset* of the ludic state transition graph can now be precisely defined: it

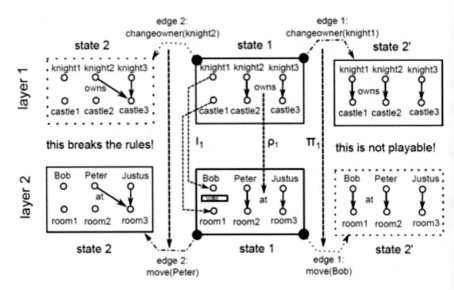

Fig. 7.3 Illustration of inconsistent state transition graphs on the narrative (*upper*) and environmental (*lower*) layer, given the rule that no two persons can own a single place

consists of exactly those edges which do not have a corresponding edge in the environmental state transition graph, and the *breakable subset* of the environmental state transition graph are exactly those transitions that do not have a corresponding edge in the ludic state transition graph.

In order to formally capture this idea, we define a couple of functions which homomorphically translate between state transition graphs on different layers:

$$f^i_{fact} = (a, R_{ij}, b) := (\iota_i(a), \rho_i(R_{ij}), \iota_i(b)) \text{(translate facts between layers)}$$

$$fset(\{x,...,z\}) := \{f(x),...,f(z)\} \text{(apply a function to a set)}$$

$$f^i_{edge}(s_1, T_{ij}, s_2) := (f^i_{fact} set(s_1), \pi_i(T_{ij}), f^i_{fact} set(s_2)) \text{(translate transition edges)}$$

Here, π_i is a mapping between state transition classes on different levels as defined above. State transition graphs have states s_{ij} as nodes (compare the squares in Fig. 7.3) and transitions between states as edges (compare edges between squares in Fig. 7.3) which are labelled by a state transition class T_{ij}. An example for such an edge would be "move(Peter)" in Fig. 7.3. f^i_{edge} translates such transition edges of a state transition graph into edges on another game level. An edge $e_0 = (s_{01}, T_{01}, s_{02})$ of a ludic state transition graph G_0 homomorphically translates to edge $e_2 = (s_{21}, T_{21}, s_{22})$ of an environmental state transition graph G_2 if states are mapped *homomorphically* and $T_{21} = \pi_1(\pi_0 T_{01})$ is a result of translating state transition classes by the mapping. A homomorphic translation of a state transition graph $G_i = (N_i, E_i,)$ therefore can be expressed by:

$$v_i(G_i) = (f^i_{fact} \, set\,(N_i), f^i_{edge} set(G_i))$$

Note that each node from N_i in this graph denotes a whole game state on layer i, and thus the graph cannot be easily visualized. In order to make state transition graphs more illustrative, we refer to the simple example given in Sect. 7.5, which contains a state transition graph on the ludic level together with possible translations (embeddings) into lower levels. We use these abstract ideas in the following to suggest novel quality criteria for breakability and playability of LBGs.

7.4.3   Quality Criteria

How can the constraints on each layer together with a mapping be used to determine quality criteria for localization?

7.4.3.1   Playability and Breakability

Ideally, an embedding is such that higher layer constraints (ludic and narrative) are precisely reflected in lower layer constraints (environment). If this is not the case, then either actions foreseen on the ludic and narrative layers are not possible in an environment (non-playability), or it becomes easy in an environment to break the rules of the game (breakability), because actions are possible which are against the rules of the game.

In order to capture these two qualities, we assume that *the space of state transitions* can be computed on each layer *independently*, based on the particular constraints of that layer. We capture these state transitions on each layer with state transition graphs $G(N, E)$, where N is the set of graph nodes and E the set of (transition-class labeled) edges between nodes:

$$G_0 = (N_0, E_0), \text{where } N_0 \subseteq \text{ possible states on ludic layer, and } E_0 \subseteq N_0 \times \Pi_0 \times N_0$$

$$G_1 = (N_1, E_1), \text{where } N_1 \subseteq \text{ possible states on narr. layer, and } E_1 \subseteq N_1 \times \Pi_1 \times N_1$$

$$G_2 = (N_2, E_2), \text{where } N_2 \subseteq \text{ possible states on env. layer, and } E_2 \subseteq N_2 \times \Pi_2 \times N_2$$

The set difference[10] between independently determined graphs on a given layer and graphs translated from higher layers is a measure for the quality of an embed-

---

[10] The operator for subtracting a set from another one is \. The set difference of two graphs $G_1 = (N_1, E_1) \backslash G_2 = (N_2, E_2)$ is defined as $N_1 \backslash N_2, E_1 \backslash E_2$.

ding, because it captures all transition possibilities caused by non-compatible constraints. Degrees of *playability* and *breakability* with respect to different layers may therefore be defined most easily in terms of the relative size[11] of the following intersections:

$$Q^0_{breakability} = \frac{|\, G_2 \setminus v_1(v_0(G_0))\,|}{|\, G_2\,|} = 1 - \frac{|\, G_2 \cap v_1(v_0(G_0))\,|}{|\, G_2\,|}$$

$$Q^0_{playability} = \frac{|\, G_2 \cap v_1(v_0(G_0))\,|}{|\, v_1(v_0(G_2))\,|}$$

$$Q^1_{breakability} = \frac{|\, G_2 \setminus v_1(G_1)\,|}{|\, G_2\,|} = 1 - \frac{|\, G_2 \cap v_1(G_1)\,|}{|\, G_2\,|}$$

$$Q^1_{playability} = \frac{|\, G_2 \cap v_1(G_1)\,|}{|\, v_1(G_1)\,|}$$

However, since graph sizes only insufficiently capture the effect on possible game strategies, it may be more adequate to measure these qualities in terms of *possible paths* from start to goal in the corresponding state transition graphs. This captures in how far possible *win strategies* are affected by constraint propagation. Suppose we denote the set of possible paths through a graph G from a start to a goal in G by the function $paths_{goal}(G)$, then:

$$Q'^0_{breakability} = 1 - \frac{|\, paths_{goal}(G_2 \cap v_1(v_0(G_0)))\,|}{|\, paths_{goal}(G_2)\,|}$$

$$Q'^0_{playability} = \frac{|\, paths_{goal}(G_2 \cap v_1(v_0(G_0)))\,|}{|\, paths_{goal}(v_1(v_0(G_0)))\,|}$$

$$Q'^1_{breakability} = 1 - \frac{|\, paths_{goal}(G_2 \cap v_1(G_1))\,|}{|\, paths_{goal}(G_2)\,|}$$

$$Q'^1_{playability} = \frac{|\, paths_{goal}(G_2 \cap v_1(G_1))\,|}{|\, paths_{goal}(v_1(G_1))\,|}$$

---

[11] Denoted by dashes around sets. The size of a graph is defined as its number of edges.

7.4.3.2 Authenticity

Authenticity describes in how far entities conceptually resemble the entities into which they are mapped. A non-authentic localization, e.g., would map places of a narrative to arbitrary places in an environment, without taking into account whether the place experience fits to the place in the narrative. For example, a medieval game may be playable in New York but the specific localization may not give rise to a very authentic experience. Even in a medieval city center, there may be more or less authentic localizations of a particular game.

In order to capture authenticity, we need to capture relevant aspects of *place experience*, such as perceptual similarity (visual, auditory, haptic qualities) and conceptual similarity (such as historical relatedness or taxonomic distance). If we can express these aspects in terms of concepts in our game ontology, then we can use existing *semantic similarity measures* in order to measure authenticity. For example, Rodriguez and Egenhofer (2004) and Janowicz (2006) proposed elaborated similarity measures for geospatial object classes. A simple kind of similarity (*sim*) between two different entities e_1 e_2 (e.g. two places or two objects) would be to measure the maximum-standardized shortest distance (*dist*) between their classes in the graph of the game ontology (*O*):

$$sim(e_1, e_2) = 1 - (\frac{dist(O, e_1, e_2)}{\max_{i,j} dist(O, e_i, e_j)})$$

Based on such a simple measure or a more elaborate one, authenticity could be defined as an aggregated similarity value:

$$Q^0_{Authenticity} = agg_{i=1}^{|D_0|} sim(e_i, \iota_1(\iota_0(e_i)))$$

$$Q^1_{Authenticity} = agg_{i=1}^{|D_1|} sim(e_i, \iota_1(e_i))$$

The aggregation function *agg* could be, e.g., a weighted sum with weights specific to the kinds of entities. Furthermore, one could also take into account similarities between ontology classes and properties as well as between corresponding state transitions into account.

7.4.3.3 Game Balancing

Another relevant but more ludic quality of any (also non-location-based) multi-player game is determined by its balancing. An unbalanced game, i.e., a game in which one player has a dominant winning strategy, will be conceived as disappointing for the loosing player, and as not very challenging for the winning player. Previous work by Schlieder et al. on Geogames has pointed out that, due to the temporal duration of the *move* action, the balancing of a LBG is particularly challenging, since running as fast as possible may easily become a dominant strategy (Schlieder et al. 2006). Though race games (e.g., **Zombies Run!** or **Can You See Me Now** (Benford et al. 2003)) may have their particular charm and motivate their players to go running, they miss the intellectual challenge of reasoning over a state space (Schlieder et al. 2006).

The spatialization of a LBG, together with the means of transportation available, influences how long it takes to locomote between places. The duration of a *move* action can only be determined after mapping it down to the environmental layer. Consequently, we identify *game balancing* as an important criterion of game localization. Tool-supported state-space analyses (Kiefer and Matyas 2005) can help simulate the spatiotemporal dynamics of a game for a given localization, yielding a numeric value that quantifies the degree of balancing. The most likely outcome of the game, as well as the number of actions each player will likely perform until that outcome is reached (given both act rationally), are two possible measures (Schlieder et al. 2006). Note that in games featuring moving (non-player) agents, the spatio-temporal balancing of a game is also influenced (and can be regulated) by the agents' speed (refer to Kiefer et al. (2005)). In general, it seems that a good balancing strategy in designing LBGs is to prevent action types which require speed from dominating the state space, e.g., by sprinkling strategic thinking actions inside a game via the game rules. We end our discussion on game balancing of LBGs here, because this problem has been extensively treated in previous work. For a game example in which balancing is of particular importance, see Sect. 7.6.

7.5 Relocalizing a Simple Conquer Game

To illustrate our quality measures, take, for example, the following simple game. Suppose there is a single player and states are described by the following vocabulary (abbreviations in brackets are used in Fig. 7.4).

$$D_0^{Places} = \{Info,\ Depot,\ Target,\ Home(H)\}$$

$$D_0^{Objects} = \{Object(O)\}$$

$$D_0^{Agents} = \{Player,\ Informant(I),\ Enemy(E)\}$$

$$\mathfrak{R}_0 = \{at(@), knows, has\}$$

$$\Pi_0 = \{move,\ ask,\ take,\ attack\}$$

The idea of this conquer game on the *ludic layer* is illustrated by 15 states (generated by according rules) in the state transition graph in Fig. 7.4: in this game, players need to find local information/equipment in an environment in order to conquer a target. At the beginning (dotted arrow on the bottom left), the player is located at

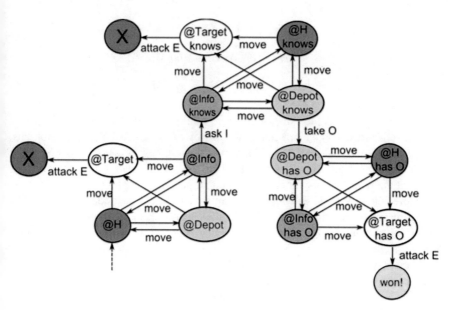

Fig. 7.4 State transition graph of a simple (single player) conquer game (ludic layer). Nodes denote states, labels in nodes denote facts about the player that are true in this state. Labels of edges denote state transition classes. Players in crossed states lose the game, and there is a single winning state

home(*H*) and can move to three other places. One of these places is the *target* that needs to be conquered to win the game. The player can directly *move* to the target, however, then lacks a resource (an object) necessary to win an *attack* of an *enemy* located at that target place, and thus will immediately loose the game (denoted by state *X*). The player thus first needs to find out where this object is located by asking a person in another place (*Info*), and once she *knows* where that object is, she can move to the *Depot* place, *take* the object *O*, move to the target, and win the attack of the enemy with the help of this object.

7.5.1   Medieval Fantasy Embedding at "Schloss Burg"

In a medieval fantasy narrative of this game, the roles may be distributed as follows (where the embedding from the ludic level is into notions at equal positions in the following listing):

$$D_1^{Places} = \{Forest, \ Cave, \ Castle, \ Village\}$$

$$D_1^{Objects} = \{Wand\}$$

$$D_1^{Agents} = \{Wizard, Dwarf, Witch\}$$

$$\mathfrak{R}_1 = \{at(@), knows, has\}$$

$$\Pi_1 = \{walk,\ ask,\ take,\ attack\}$$

Now the game tells the story (see Fig. 7.6a) of a wizard who wanders through a village and learns that an evil witch in the nearby castle has enslaved its inhabitants. The wizard promises to free the village from the reign of the witch. The way to the castle inevitably leads through a forest, where the wizard can ask a dwarf, who tells him that the witch put a spell on the castle that prevents people from escaping, and therefore can only be defeated using a magic wand, which is hidden in a cave. The wizard needs to find the wand and enter the castle to keep his promise. Note that under this embedding, all states of the game reappear homomorphically (see Fig. 7.6a), however, some state transitions were removed to streamline the story (e.g., there is no possibility to return to the village after a certain point in the story).

Suppose we furthermore embed this narrative into the environment of a real castle, such as "Schloss Burg"[12] in Germany (see Fig. 7.5). The role of the village could be played by *"Unterburg"*, which is part of a small town (Burg an der Wupper) located directly at the foot of the hill on top of which the castle is located, the forest could be played by *"Schlossberg"*, the woody hill slope through which a footpath leads to the top, and the cave could be embodied by a *playground* beneath the castle. The sphere of influence of the castle could involve a narrow buffer or boundary surrounding the castle (compare Fig. 7.5):

$$D_2^{Places} = \{Schlossberg,\ Playground,\ SchlossBurg,\ Unterburg\}$$

$$D_2^{Objects} = \{Wand_{virt}\}$$

$$D_2^{Agents} = \{Player,\ Dwarf_{virt},\ Witch_{virt}\}$$

$$\mathfrak{R}_2 = \{at(@),\ knows,\ has\}$$

$$\Pi_2 = \{walk,\ ask,\ take_{virt},\ attack_{virt}\}$$

Note how some of the entities and actions are *virtual* (the witch, the dwarf and the wand), while others correspond to things in physical reality.

Note furthermore the state transition differences imposed by *environmental affordances* (compare Figs. 7.6b and 7.5a): under this embedding, certain direct walks, namely the ones between the "forest" (Schlossberg) and the "cave" (Playground) are not possible anymore, because the footpath (black dotted line in Fig. 7.5a) through the forest inevitably leads to the castle first. Furthermore, a player

---

[12] https://en.wikipedia.org/wiki/Burg_Castle_(Solingen).

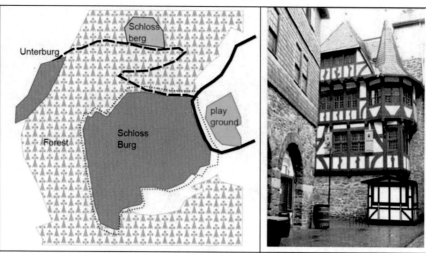

(a) Places around Schloss Burg.   (b) Schloss Burg a.d. Wupper,
                                                         Germany (CC BY-SA 2.0[12] cour-
                                                         tesy by Plybert 49 on Flickr)

Fig. 7.5 The environment for embedding the medieval narrative. (**a**) Places around Schloss Burg.
(**b**) Schloss Burg a.d. Wupper, Germany (CC BY-SA 2.0 (https://creativecommons.
org/licenses/by-sa/2.0/) courtesy by "Polybert49" on Flickr)

can leave the castle and get to the playground by taking the footpath leading past the
castle's exterior wall. The latter breaks the rules of the game, whereas the former
renders the game unplayable under this embedding. To be more precise, Table 7.2
shows the exact numbers for playability and breakability as defined in this chapter
together with the underlying graph-based measures regarding the medieval
embedding.

Note that only the path-based measures (Q') actually reveal that the game is practically unplayable under this embedding (*playability* = 0), and that every possible
strategy will break the ludic as well as narrative rules of the game (*breakability* = 1).
Note also that playability and breakability are not simply $(1 - x)$ of each other.

7.5.2   *Crime Story Embedding in "Little Italy"*

Suppose we embed the medieval fantasy narrative ($n1$ in Table 7.3) into an urban
environment, such as the Little Italy district in New York ($e2$ in Table 7.3). For
instance, the narrative "Village" would map to "Angelo's" (an Italian restaurant),
the medieval "Forest" to "Ravenite Social Club", etc. (refer to Table 7.3). The
authenticity of this embedding, taking into account ontological differences between
classes and properties, should be rather low.

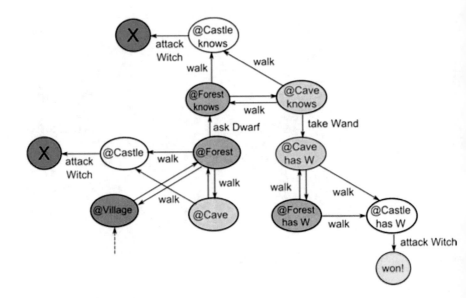

(a) State transition graph of the medieval fantasy narrative.

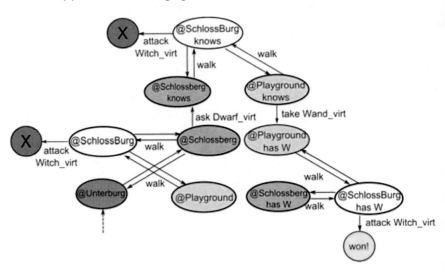

(b) State transition graph of the game environment of Schloss Burg.

Fig. 7.6 State transition graphs on narrative (**a**) and environmental (**b**) layers for the medieval fantasy embedding. (**a**) State transition graph of the medieval fantasy narrative. (**b**) State transition graph of the game environment of Schloss Burg

Table 7.2 Playability and breakability measures for the medieval embedding

	$|G_2|$	$|vG|$	$|G_2 \cap vG|$	Breakability	Playability
Q^0	19	31	11	0.421	0.355
Q^1	19	19	13	0.316	0.684
Q^{10}	1	20	0	1	0
Q^{11}	1	2	0	1	0

$|G_2|$ = cardinality of state transition graph on environmental layer, $|vG|$ = cardinality of translation from 0/1 layer to environmental layer, $|G_2 \cap vG|$ cardinality of intersection. Cardinalities are either of edge sets (Q) or of sets of start-goal paths (Q')

Table 7.3 Authenticity for two narrative and two environmental embeddings

Ludic	Narrative 1 (n1)	Narrative 2 (n2)	Environment 1 (e1)	Environment 2 (e2)	Similarities n1,e1	n1,e2	n2,e2
Home	Village	Restaurant	Unterburg	Angelo's	0.75	0.25	1
Info	Forest	Nightclub	Schlossberg	Ravenite Social Club	1	0.25	1
Depot	Cave	Rifle Store	Playground	John Jovino Gun Shop	0.5	0.25	1
Target	Castle	Bank	Schloss Burg	City Bank	1	0.5	1
$Q^1_{Authenticity}$					0.8125	0.3125	1

Figure 7.7 displays a simple ontology of the different types of places, which we can use to measure authenticity. And in fact, based on this ontology, it turns out that the averaged similarity (as defined in Sect. 7.4.3) over all places is 0.31 (see Table 7.3).

For the New York environment, a different narrative would provide better authenticity values, and thus a better gaming experience. Consider a narrative playing in the times of the Mafia of the 1920s ($n2$ in Table 7.3): the player is member of a Mafia family, seated at some restaurant, and has the goal of robbing a Bank. For this, he needs to move to a nightclub to find out how to get a gun. Some other Mafioso in the nightclub tells him to rob a specific gun shop. This embedding yields an averaged similarity of 1, since all places match places of identical classes (Table 7.3). Note that the original medieval fantasy embedding at Schloss Burg yields also a high authenticity value of 0.81, which is a bit lower than 1 because the playground is not an ideal place for the role of the cave.

7.6 Localization of an Existing Multi-player Game: CityPoker

Here we demonstrate how relocalization can be applied to an existing game: **CityPoker**, a multi-player LBG introduced in (Kiefer et al. 2005; Kremer et al. 2013). As for any serious game, its state transition graph is too complex to be

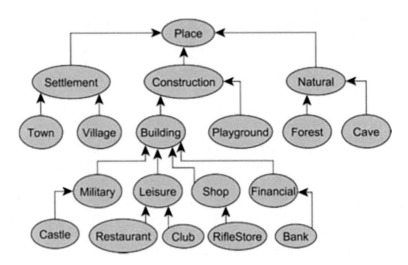

Fig. 7.7 A hierarchy of place types used for measuring authenticity based on semantic similarity

visualized. We provide a simplified rule description here; the extended rule set can be found in Kiefer et al. (2005).

In **CityPoker**, two players each aim at improving their hand of five cards by exchanging these with cards hidden in the environment. There are 20 cards in the game ($\{\clubsuit, \diamondsuit, \heartsuit, \spadesuit\} \times \{10, J, Q, K, A\}$), 10 of which are on the players' hands, and 10 hidden in five caches. Players can exchange at most once at each cache, which means they drop one card and pick another. Figure 7.8 (left) illustrates a possible initial card distribution for Bamberg, Germany, as well as the players' starting positions. The end evaluation follows that of the traditional Poker game (Royal Flush > Four of a kind > ... > One Pair).

The ludic level contains the following things:

$$D_0^{Places} = \{place_1, \ldots, place_5, \, start_1, \, start_2\}$$

$$D_0^{Objects} = \{item_1, \ldots, \, item_{20}\}$$

$$D_0^{Agents} = \{player_1, player_2\}$$

$$\Re_0 = \{at(@), has\}$$

$$\Pi_0 = \{move, exchange\}$$

where the mechanics of the game are modeled with a large state graph describing all possible sequences of moving and exchanging cards. There is a trivial bijective mapping from the ludic to the narrative level (similar for state graphs):

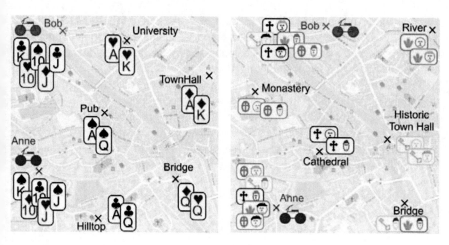

Fig. 7.8 CityPoker with two different narrative and environmental embeddings in Bamberg, Germany (*left*: original, *right*: medieval version; basemap: OpenStreetMap)

$$D_1^{Places} = \{cache_1, \dots, cache_5, start_1, start_2\}$$

$$D_1^{Objects} = \{heart10, heartJ, \dots, spadesA\}$$

$$D_1^{Agents} = \{pokerplayer_1, pokerplayer_2\}$$

$$\mathfrak{R}_1 = \{hasSelectedCache, hasOnHand\}$$

$$\Pi_1 = \{selectCache, swapCard\}$$

The localization displayed in Fig. 7.8 (left) yields in the following sets of game elements on the environmental level:

$$D_2^{Places} = \{TownHall, Bridge, start_1, start_2\}$$

$$D_2^{Objects} = \{heart10_{virt}, heartJ_{virt}, \dots, spadesA_{virt}\}$$

$$D_2^{Agents} = \{Bob, Anne\}$$

$$\mathfrak{R}_2 = \{at(@), has_{virt}\}$$

$$\Pi_2 = \{bicycle, keyPress\}$$

Let us assume our localization allows for locomotion by bicycle between each pair of caches, and the game software ensures that cards can only be exchanged following the ludic rules. In that case, the localization is perfectly playable and not at all breakable. Authenticity, however, is rather weak (none of the selected places is associated with Poker), and a relocalization within Bamberg would not help either: the historical center of Bamberg is characterized by medieval buildings and tourist attractions, not with a single gambling place.

This can be solved by changing the narrative to a medieval setting, while keeping the ludic rules fixed: the four colors ($\{\clubsuit, \blacklozenge, \heartsuit, \spadesuit\}$) could be replaced by four competing parties that were relevant in medieval times: {*CatholicChurch, Benedictines, Citizens, Peasants*}. For each party, we could select five professions replacing the Poker numbers, such as {*Abbot, Vice-Abbot, Treasurer, Cellarer, Monk*} for Benedictines, and {*Major, Vice-Major, Merchant, Blacksmith, Worker*} for Citizens:

$$D_{I'}^{Places} = \{cathedral, \ldots, monastery, start_1, start_2\}$$

$$D_{I'}^{Objects} = \{BenedictineAbbot, BenedictineViceAbbot, \ldots, CitizenWorker\}$$

$$D_{I'}^{Agents} = \{delegate_1, delegate_2\}$$

$$\Re_{I'} = \{at(@), hasAsFollower\}$$

$$\Pi_{I'} = \{horseRiding, convinceFollower\}$$

In this narrative, players are delegates on some diplomatic mission with the goal of convincing influential people (which are considered items here, not agents). The winning condition is defined in a way consistent with Poker: all from one party > four of the same level > … > two of the same level. It is now possible to find a localization in Bamberg which ensures high authenticity (e.g., Bamberg Cathedral, Michaelsberg monastery, etc.; see Fig. 7.8). Finally, it will most likely be necessary to change the two start positions to keep the game dynamics balanced, which is out of the scope of this chapter.

7.7 Discussion and Conclusion

Based on a layered model of game localization, we have suggested novel measures for playability, breakability and authenticity of possible environmental embeddings of a game. Since our approach involves game narratives, it takes into account some of the "play" aspect of games. It also contributes to the challenge of "deep" localization of games, which goes beyond superficial spatialization to consider embedding of games into places and possible actions (affordances).

Now, one game embedding can be compared with another one in order to determine the optimal one given an environment. This provides a way to answer research questions 1 and 2 of section 1 about localization and re-localization, as it gives us novel and relevant criteria to evaluate possible localizations with respect to narratives, roles and environmental affordances. However, in this chapter, we have not yet addressed the problem of *searching* for good or optimal localizations. Based on future research, it might also become possible to search for a game that has the highest quality of embedding into some given environment, addressing question 3 (gamification) of section 1.

Here are a number of open research questions that need to be addressed in order to reach these goals.

First, to what extent does our approach really capture meaningfulness and the play aspect of games? In how far could it be used for *meaningful gamification* of environments? The existing research on meaningful gamification is in a very early stage (Nicholson 2012; Hassenzahl and Laschke 2015), which means it is open what aspects of gaming activity really need to be taken into account. We think our chapter gives some suggestions on what criteria might be relevant.

Second, our approach requires that state formalizations and state transition graphs are present on all game layers. Which sensors/observations are needed in order to generate state transitions on the environmental layer? How can we formalize state transition constraints on ludic as well as narrative layers? How can we compute state transition graphs given constraints? This can be of different complexity, depending on the nature of these constraints.

Third, and most importantly, the computation of the localization quality of a given embedding, as well as the search for an optimal embedding given an environment are both computationally complex. Computing playability and breakability in a strategic manner requires computing all start to goal paths in transition graphs on all three layers. Computing authenticity requires similarity computations for each mapped symbol. Searching over the set of possible game localizations given states on two layers is a combinatoric problem $\binom{n}{m}$ where n is the size of the union of domain, relation and state transition symbols on one layer and m on the other. However, the latter problem can always be simplified by certain practical considerations, such as a fixed start location of a user, and a restricted relation or state transition mapping.

References

Ahlqvist O, Loffing T, Ramanathan J, Kocher A (2012) Geospatial human-environment simulation through integration of massive multiplayer online games and geographic information systems. Trans GIS 16(3):331–350

Rodriguez MA, Egenhofer MJ (2004) Comparing geospatial entity classes: an asymmetric and context-dependent similarity measure. Int J Geogr Inf Sci 18(3):229–256

Benford S, Anastasi R, Flintham M, Drozd A, Crabtree A, Greenhalgh C, Tandavanitj N, Adams M, Row-Farr J (2003) Coping with uncertainty in a location-based game. IEEE Pervasive Comput 2(3):34–41

Benford S, Magerkurth C, Ljungstrand P (2005) Bridging the physical and digital in pervasive gaming. Commun ACM 48(3):54–57

Bichard J, Brunnberg L, Combetto M, Gustafsson A, Juhlin O (2006) Backseat playgrounds: pervasive storytelling in vast location based games. In: Entertainment computing-ICEC 2006, Springer, pp 117–122

Cresswell T (2013) Place: a short introduction. Wiley, New York
Deterding S, Dixon D, Khaled R, Nacke L (2011) From game design elements to gamefulness: defining gamification. In: Proceedings of the 15th international academic MindTrek conference: envisioning future media environments, ACM, pp 9–15
Dourish P (2006) Re-space-ing place: place and space ten years on. In: Proceedings of the 2006 20th anniversary conference on computer supported co-operative work, ACM, pp 299–308
Gangemi A, Presutti V (2010) Towards a pattern science for the semantic web. Semant Web 1(1-2):61–68
Gibson JJ (2013) The ecological approach to visual perception. Psychology Press, Boston
Hajarnis S, Headrick B, Ferguson A, Riedl MO (2011) Scaling mobile alternate reality games with geo-location translation. In: Interactive storytelling, Springer, New York, pp 278–283
Hassenzahl M, Laschke M (2015) Pleasurable troublemakers. The gameful world: approaches, issues, applications. MIT Press, Cambridge, MA, p 167
Hawkins J, Blakeslee S (2007) On intelligence. Macmillan, New York
Hinske S, Lampe M, Magerkurth C, Röcker C (2007) Classifying pervasive games: on pervasive computing and mixed reality. In: Magerkurth C, Röcker C (eds) Concepts and technologies for pervasive games—a reader for pervasive gaming research, vol 1. Shaker Verlag, Aachen, p 20
Janowicz K (2006) Sim-DL: Towards a semantic similarity measurement theory for the description logic ALCNR in geographic information retrieval. In: On the move to meaningful internet systems 2006: OTM 2006 workshops, Springer, pp 1681–1692
Jenkins H (2004) Game design as narrative architecture. Computer 44(3):118–130
Kiefer P, Matyas S (2005) The geogames tool: balancing spatio-temporal design parameters in location-based games. In: Mehdi Q, Gough N (eds) Proceedings of the 7th international conference on computer games: artificial intelligence, animation, mobile, educational and serious games (CGAMES 2005), Angoulême, France, pp 216–222
Kiefer P, Matyas S, Schlieder C (2005) State space analysis as a tool in the design of a smart opponent for a location-based game. In: Masuch M, Hartmann K, Beckhaus S, Spierling U, Sliwka F (eds) Proceedings of the Computer Science and Magic 2005, GC Developer Conference Science Track, Messe Leipzig, Leipzig, Germany, ISBN:3-9804874-3-1
Kiefer P, Matyas S, Schlieder C (2007) Playing location-based games on geographically distributed game boards. In: Magerkurth C, Akesson KP, Bernhaupt R, Björk S, Lindt I, Ljungstrand P, Waern A (eds) 4th international sym posium on pervasive gaming applications (PerGames 2007). Shaker Verlag, Aachen, Salzburg, Austria
Kremer D, Schlieder C, Feulner B, Ohl U (2013) Spatial choices in an educational geogame. In: Games and virtual worlds for serious applications (VS-GAMES), 2013 5th international conference on, IEEE, pp 1–4
Lemos A (2011) Pervasive computer games and processes of spatialization: informational territories and mobile technologies. Can J Commun 36(2):277–294
MacLennan B (2009) Analog computation. In: Meyers RA (ed) Encyclopedia of complexity and systems science. Springer, New York, pp 271–294. https://doi.org/10.1007/978-0-387-30440-3
Montello DR (1993) Scale and multiple psychologies of space. In: Spatial information theory—a theoretical basis for GIS. Springer, New York, pp 312–321
Montola M (2005) Exploring the edge of the magic circle: defining pervasive games. In: Proceedings of DAC, vol 1966, p 103
Nicholson S (2012) A user-centered theoretical framework for meaningful gamification. In: Games+Learning+Society 8, Madison, WI
Nicklas D, Pfisterer C, Mitschang B (2001) Towards location-based games. In: Proceedings of the international conference on applications and development of computer games in the 21st century: ADCOG, vol 21, pp 61–67
Paay J, Kjeldskov J, Christensen A, Ibsen A, Jensen D, Nielsen G, Vutborg R (2008) Location-based storytelling in the urban environment. In: Proceedings of the 20th Australasian conference on computer-human interaction: designing for habitus and habitat, ACM, pp 122–129

Petruck MR (1996) Frame semantics. In: Verschueren J, Östman J-O, Blommaert J, Bulcaen C (eds) Handbook of pragmatics. John Benjamins, Philadelphia, pp 1–13

Reid J (2008) Design for coincidence: incorporating real world artifacts in location based games. In: Proceedings of the 3rd international conference on digital interactive media in entertainment and arts, ACM, pp 18–25

Scheider S (2012) Grounding geographic information in perceptual operations, vol 244. IOS Press, Amsterdam

Scheider S, Janowicz K (2010) Places as media of containment. In: Proceedings of the 6th international conference on geographic information science (extended abstract)

Scheider S, Janowicz K (2014) Place reference systems: a constructive activity model of reference to places. Appl Ontol 9(2):97–127

Scheider S, Kiefer P, Weiser P, Raubal M, Sailer C (2015) Score design for meaningful gamification. In: Researching gamification: strategies, opportunities, challenges, ethics, Workshop at CHI 2015

Schlieder C (2014) Geogames—Gestaltungsaufgaben und geoinformatische Lösungsansätze. Informatik-Spektrum 37(6):567–574

Schlieder C, Kiefer P, Matyas S (2006) Geogames: designing location-based games from classic board games. IEEE Intell Syst 21(5):40–46

Seamon D (1979) A geography of the lifeworld. St. Martin's Press, New York

Searle JR (1995) The construction of social reality. Simon and Schuster, New York

de Souza e Silva A (2008) Hybrid reality and location-based gaming: redefining mobility and game spaces in urban environments. Simul Gaming 40(3):404–424

Stenros J (2015) Behind games: Playful mindsets and transformative practices. In: Walz SP (ed) The gameful world: approaches, issues, applications. MIT Press, Cambridge, MA, p 201

Tuan YF (1977) Space and place: the perspective of experience. University of Minnesota Press, Minneapolis

Tuan YF (1991) Language and the making of place: a narrative-descriptive approach. Ann Assoc Am Geogr 81(4):684–696

Walther BK (2005) Atomic actions–molecular experience: theory of pervasive gaming. Comput Entertain (CIE) 3(3):4–4

Chapter 8
The Design and Play of Geogames as Place-Based Education

Jim Mathews and Christopher Holden

8.1 Introduction

A key affordance of games and technologies that emphasize geolocativity is their ability to bring people to new places, and mobilize aspects of those places to facilitate playful and interactive experiences. At a basic level, we can think of a mobile application's ability to locate where you are (via GPS, QR codes, Bluetooth beacons, image recognition, etc.) and use that information to direct you towards specific objects or locations within a place or encourage particular types of interactions. Not only where you are, but *what you do* in specific locations (e.g. take and share photos) can become part of a mediated experience that encourages you to act and interact differently in the world. Games that take into account a person's physical location can be single or multiplayer: relying on players' co-presence in a given place either synchronously or asynchronously, or designed to elicit interactions among users in different places, where each player contributes based on the unique features of their location. In essence, mobile technologies can provide new ways of knowing *where you are*: from highlighting your position in time and space to helping you identify and interpret your surroundings, to cultivating new sociocultural and identity based modes of awareness. Geogames can amplify these possibilities by providing the conditions (e.g., the context and inspiration) needed to encourage players to take action within a particular place. As a result, game authors can facilitate new ways of seeing the world, open new modes of access to the worlds that players already encounter, and create new worlds and narratives layered on top of existing reality.

J. Mathews (✉)
Field Day Lab, University of Wisconsin, Madison, WI, USA
e-mail: jmmathews@gmail.com

C. Holden
University of New Mexico, Albuquerque, NM, USA

© Springer International Publishing Switzerland 2018
O. Ahlqvist, C. Schlieder (eds.), *Geogames and Geoplay*, Advances in
Geographic Information Science, https://doi.org/10.1007/978-3-319-22774-0_8

As with other media, it is possible for instructors and researchers to take geogames originally designed for entertainment purposes and use them within educational contexts. But, thanks to the emergence of new authoring tools (e.g., platforms such as **ARIS** and **TaleBlazer** that allow non-programmers to build location-based media) it is also increasingly possible for these same parties to design geogames specifically with educational goals in mind. It is these education-oriented games, rather than commercial ones, that we focus on in this chapter.

Mediating location-based learning using mobile devices is not a new idea. Even before advanced mobile technologies like smartphones were in wide use, researchers were exploring the basic affordances and educational possibilities of geogames to help situate content within specific locations and contexts. While this approach has been diversely enacted, it typically involves using geo-locative games to develop students' understanding of the academic language, concepts and skills associated within a specific domain through structured, but "real world," problems and cases. Two early examples of this approach include **Environmental Detectives** (Klopfer and Squire 2004) and **Mad City Mystery** (Squire and Jan 2007). Each of these games requires players to assume professional roles, visit local places, and collect and analyze data to solve a domain-specific problem. The premise of this type of game, is that the authenticity, concreteness and immediacy of local place combine to aid learning, by providing concrete instances and applications of abstract models and concepts. Game mechanics used within these scenarios employed verbs beyond "recall" (e.g., collect, interview, observe, interpret), and roles (e.g., scientist, health official, environmental historian, medical doctor), to facilitate players' interactions within a given domain (Squire et al. 2007). However, while they were played outdoors and incorporated authentic roles and problems, the games (and the people who played them), were often "dropped onto" places without taking into account their unique cultural or ecological characteristics.

Ongoing design work in this area has added to the diversity of contexts in which this type of content-centric geogame has been used to support specific learning goals, and has covered a lot of ground in terms of settings (school, after-school, summer, museum-based, etc.), subject areas (science, language arts, foreign language instruction, history, ornithology, environmental science, civics, math, etc.), and ages (Dikkers et al. 2012; Holden et al. 2015). Because this line of research is still in its infancy, there is a need to further explore and develop mechanics, subjects, game types, etc. that produce good examples of how place and content can come together in harmony. This approach of using geoplay to help situate academic skills and content leaves room for further consideration of how the same basic affordances of mobile media might be used to investigate place in other ways. In reflecting on past designs and planning for the future, it is useful to ask questions about the relationship between particular games and the place(s) they are played: Is the game situated in place merely in a mechanical sense, or does it actually integrate with the historical, social, or other particulars of a place? Is the situatedness of the game designed solely to improve the acquisition of academic knowledge and skills, or does play actually connect players more deeply to the environments in which they live?

Asking these questions ourselves over the years has produced a profound shift in what we emphasize most in our games. In our current work, we tend to design (and help others design) geogames and associated activities that focus on place and its consequents, rather than academic subject areas. In part, a move that has been influenced by our shared interest in place-based education. Making place the subject of our games, instead of simply the method, has numerous implications for what they look like and the roles they play in the lives of their creators and players.

The rest of this chapter articulates why we suggest a more thorough adoption of place-based education in the design and use of geogames and geoplay for learning, and presents some of our experiences pursuing this goal in our own work over the last several years. It ends with suggestions for how we can invite others to engage in their own design experiments in this area, particularly within the context of building a larger and more diverse group of game designers.

8.2 What Is Place-Based Education and Why Should Educators Care About It?

Despite having deep roots in environmental education, the "big tent" of place-based education has evolved over time to include a range of teaching approaches, including cultural journalism, ecological education, local entrepreneurialism, and service learning (Gruenewald 2003; Orr 1994; Smith and Sobel 2010). At the center of each of these approaches, however, rests a commitment to situating students' learning experiences within the local community. It is important to note, that while place-based education emphasizes the exploration of *the local*, it also provides methodological and values-based guidance for how to enact local studies. As such, a place-based pedagogy not only speaks to *what* should be studied, but also *how* and *why*.

For this chapter we have selected a number of key components of place-based education that are relevant to the broader discussion of geogames and geoplay for learning.

- *Locally-focused*: Place-based education emphasizes the study of local cultural and ecological systems, as well as the relationships between them. Curricular goals often emerge from local needs and contexts rather than centralized standards.
- *Authentic questions, issues, challenges*: Studying local systems can contextualize students' learning around "real world" problems, issues and questions. Because it foregrounds engaging learners in studying places and ideas that are relevant to them, place-based education helps students make connections between new concepts and their own lived experiences of place.
- *Interdisciplinarity*: Subject matter content and other disciplinary constructs that typically structure domain-centric inquiry take a back seat, allowing contributions from many perspectives. The particular approaches taken, as well as the

content and concepts students engage with, emerge from the particular problem, place or issue under investigation and can shift based on the nature of the inquiry or design space. Also, place-based education values local knowledge and ways of knowing under the umbrella of interdisciplinarity.

- *Focus on knowledge building*: Place-based education positions learners as producers of new knowledge, not mere consumers of information. It also emphasizes the cultivation of learning communities that extend beyond the classroom walls and build on the work generated by previous students.
- *Action oriented*: Place-based education actively engages students in co-designing learning experiences and shaping their worlds through their actions. Learning entails creating and maintaining relationships among actors, and emphasizes participation in the present rather than learning solely for an unspecified or hypothetical future.
- *Integrated Tools*: Tools are embedded in larger practices, questions, and explorations. Just as an academic content area does not find itself at the center, neither do any of the tools or artifacts used in the processes of place-based education.

When viewed holistically, it becomes clear that place-based education entails more than simply using the community as a repository for content. Instead, it represents a pedagogy that challenges the teacher-driven, content-centric, and subject-oriented approaches that dominate traditional forms of education. From another angle, and importantly for how we imagine large numbers of educators adopting this pedagogical approach, we see place-based education as a natural outcome of dissatisfaction with the ability of existing curricula to connect to learners' lives or make use of their skills outside the classroom.

As teachers we support our own students in studying local places—including their homes, neighborhoods, and nearby natural areas—as they engage with a wide range of topics, including architecture and urban design, folklore, public art, citizen science, indigenous language use, etc. In each of these instances, our students interact with local people and places, conduct fieldwork, and produce new content and knowledge through original research and design projects. Typically, this work focuses on studying both cultural and ecological systems, as well as the relationships between them. Rather than simply requiring them to remember a series of facts, this combination of community-based research and design helps students develop an understanding of how local systems operate, interact, and respond to change.

A key consideration in this work is providing learners with scaffolding that helps them explore the openness of the world without losing the saliency of focused instruction. In part, this has led to our strong interest in exploring how geo-locative games, play and narrative might be used to support and amplify place-based learning. For example, how might geogames be used to spark curiosity and help learners develop new questions and inquiry projects related to local places? How might they help learners experience and see their community from new perspectives? How might they help students engage with and think more critically about their local community(-ies)?

8.2.1 Moving from Place-Based Education to Geo-games and Geoplay

We have already described how the affordances of mobile technologies can push designers towards creating games that employ geo-locativity. However, given all of the options for enacting place-based education, why should someone committed to this pedagogical approach consider geogames and geoplay? In part, this question stems from an acknowledgment that there is not an automatic harmony between advocates of place-based education and mobile technologies and digital games. At a basic level, because of its roots in environmental and outdoor educational philosophies, as well as a critique of how digital technologies have previously been used in education, place-based educators may be especially cautious of adopting new digital technologies, especially in outdoor contexts (Gruenewald and Smith 2008). For example, it is not unusual for outdoor education programs to be organized around the goal of getting kids off of screens, at least for a bit.

Rather than seeing the capacity of technology to separate us from our surroundings, we believe there is much to be gained from exploring the potential of mobile and geo-locative technologies to bring us physically outside the classroom, provide new capacities for investigating local places, and expose aspects of the world we might typically ignore. In some cases, technology can also increase interest and build trust among participant groups when enacting place-based education. Youth may not typically identify with the types of inquiry and reflection about their local community that can feel so crucial to the educators who push for place-based education. But, providing opportunities for learners to leverage their interests and experiences using digital technologies to engage in creative exploration and production may provide an entry point for engaging students and encouraging critical inquiry (Mathews 2010; Squire and Dikkers 2012).

8.2.2 Geoplay and Geo-games for Supporting Place-based Learning

While there are many methods for implementing a place-based approach, we are intrigued by the potential of geogames and geoplay to unlock new opportunities in this area. Geogames often take advantage of physical space and place to facilitate playful interactions and experiences. From taking on new roles that mediate how players engage with places, to enabling them to enact narratives—whereby stories are not passively told, but actively lived by players—geogames can slowly introduce learners to new places, or help them see the familiar in new ways. In this section, we discuss how this potential aligns with, and can be used to support, the values and goals of place-based education. In doing so, we draw from our own work, as well as the broader field of geoplay. While we present examples of place-conscious games here, it is important to note that they represent only one type of experience students would typically engage in as part of a deeper and longer trajectory involving local studies. That is, in most cases, the games were designed in coordination with larger learning goals; for example as a way to generate questions about local places, spark interest in a particular topic or theme, or help players identify community resources.

8.2.3   Geogames and Geoplay Can Encourage Learners to Access Local Places They Don't Typically Visit

There are often aspects of our communities, despite their physical proximity, that are unfamiliar to us. In some instances, this unfamiliarity is simply the result of not having a context (e.g., a need or relational connection) for interacting within a particular place. Geogames can help address this disconnect, by providing an impetus for exploring new places and scaffolding to help players know what to do there. To get a feeling for this, let's begin by referencing a simple game that is quite popular and easy to play; hide-and-seek. Imagine playing a version of this game during a visit to an acquaintance's house. 10, 9, 8,… quick, where are you going to hide? How about behind the sofa, in a closet, or under a bed? Note that using a simple verb or mechanic like *hide*, instantly changes how you interact with the space. That is, in most circumstances it would be highly unusual for you to crawl under a bed or open a closet when visiting someone. But not so strange when playing a game that explicitly encourages you to uncover secret or hidden places. In this way, hide-and-seek exemplifies how games can combine rules and mechanics in a way that enables and leads players to access new places or places that might typically be off-limits from a cultural, or daily practices standpoint. In this regard, games and play can provide an invitation to go to a new place, or engage in an activity that might run counter to how one typically acts there.

These same principles can be applied in educational contexts. For example, our colleague John Martin developed a game called **Mystery Trip** that was played by campers at a residential outdoor camp in the woodlands of Maine. In part, it was designed to foster student, rather than instructor directed exploration by campers, and facilitate and scaffold independent discovery. In the game, which was played on mobile devices, players were required to venture off the manicured trials to avoid detection from competing teams. As a result, they were rewarded for creating their own routes through the woods, rather than using the official trail. This shift altered their relationship with the woods by changing their perspective, requiring new behaviors, and encouraging them to ask new questions (Martin 2009). In part, it achieved this goal by providing a narrative and quest structure that supplied an impetus for engaging in these new behaviors. It also provided a play space that bounded (which helped the players feel safe), but also sanctioned their exploration.

Similarly, In **Mentira**, a Spanish language game developed by Julie Sykes and Chris, players travel to a neighborhood in Albuquerque, New Mexico—one they have likely passed through but would not recognize as a neighborhood or place distinct from the city as a whole—where for an hour or two, they are embedded in a hybrid world that mixes a fictional narrative with real world locations (Holden and Sykes 2012). In addition to leading them to new places, **Mentira** requires players to use their understanding of Spanish to successfully complete quests and solve an overarching mystery tied to the unique topography and cultural history of the neighborhood. The game, which is used in the fourth semester of a Spanish sequence at the University of New Mexico, coincides with a curricular goal of making the Spanish language a part of students' lives outside the classroom. While Spanish is widely used in Albuquerque, most students in the class, though often natives of the city, do not use Spanish as part of their everyday lives. The goal of the game is not to introduce new lexical items or test students on information presented in class, but rather to give them a situation where their Spanish is something that enables them to act in new places and contexts. While, a couple of hours talking to virtual characters out in a local neighborhood is just a beginning, the game provides an entry point for the students—a bridging experience that requires them to use their Spanish in a situated manner and gives them a context for visiting a new area in their city. One can imagine building on this initial game to create a series of similar adventures, or as a way to get students started building their own Spanish and neighborhood-based games for others to play.

Games can provide a framework of meaning that connects players to new places, and gives them a purpose within them. While many games provide only a brief exposure to a place, a first hand experience—even one of short duration—can support the development of deeper understanding of, and curiosity about, a place. A basic embodied familiarity is often a key ingredient and perhaps an initial entry point to understanding a place, and something that locative games can not only help with, but improve upon over more typical experiences such as tours and introductory field surveys.

8.2.4 Geogames Can Foster New Ways of Seeing and Experiencing Places

Even in cases where we visit a place on a regular basis, games can be used to alter how we typically interact there. In other words, games can defamiliarize us with places. They can give us pause, break us out of our routines, and encourage us to look at the world anew. Again, hide-and-seek exemplifies this point. You might visit a friend's house on a regular basis, but once you are tasked with hiding, or looking for other people who are hiding, you are required to reread the landscape. A closet is no longer a place to hang coats, but somewhere to hide.

Experiencing a place differently often involves seeing it through new lenses. Places and "real world" problem spaces are inherently complex and students often need support in knowing what to look for and how to look for it. It is a nontrivial skill, for example, to conduct field observations, like those required as part of an introductory field course in geology or sociology. Games can help in this regard, by providing anchoring experiences, roles, embedded tools, and mechanics that direct players attention to various aspects of the landscape; thereby helping them identify patterns and make sense of the messiness of places.

Geogames can be used to highlight both the cultural and ecological aspects of the local places. For example, history-based games, such as **Dow Day** (Mathews 2009; Mathews and Squire 2009) and Owen Gottlieb's **Jewish Time Jump** (Rosenkrantz 2014) engage players with historical sites and events, and explore larger historical constructs such as perspective recognition and evidence-based argumentation, while games like **Digital Graffiti Gallery** (Holden 2015) scaffold cultural fieldwork, and games like **WeBird** (an ornithology game) (UW Mobile n.d.-a) help players locate, identify and document local wildlife. Other games, such as **SustainableU** (a game about energy conservation that combines field observations and mini-game based tutorials) (UW Mobile n.d.-b) and **To Pave or Not to Pave** (Mathews 2010) combine both ecological and cultural aspects of particular places and explore the relationships and tensions between them.

One way to help players experience and see familiar places in new ways is through narrative- and role-based play. In games that employ these features, players take on roles and enact narratives that help them re-inhabit the world, often in a playful manner. Narratives, particularly ones that emphasize contestation and multiple perspectives, can be used to highlight the multiplicity of place, and encourage players to seek out new perspectives on issues. For example, in **Riverside**, a game we developed that takes place in an area of Milwaukee undergoing gentrification, players are tasked with exploring a green space and making recommendations for how, and if, it should be developed. To help them formulate a proposal, players interview a series of virtual characters representing a variety of perspectives on what should be done with the land and why. The players also explore scientific data and experience how the area has changed over time, both culturally and ecologically, by viewing historical images, maps and videos, making field observations, and talking to people who live and work there. The people they meet, and the data they collect, are tied to their role as either an environmental historian or wildlife ecologist. Having separate roles in the game helps players view the place and problem space from different perspectives. It also encourages them to share information in order to build a more complete narrative.

In another game, **Up River** (Wagler and Mathews 2012), players explore a large freshwater estuary by completing a series of quests that take them to several locations seldom visited by outsiders, including an industrial area and a reclaimed wetland. In this case, the narrative is designed so that players literally see the watershed from different perspectives. They also investigate how it is used by a variety of stakeholders, including those who hold differing opinions about recreation, conservation, and development. In the game, the players interact with historical characters, who share stories and primary documents (e.g., photos, maps, journals) depicting what the area looked like in the past, as well as current residents who share stories about their own experiences living in the watershed. These stories are combined with scientific data in a way that helps players make connections between the natural environment and cultural uses (both past and present) of the land. The game also serves as a launching point for engaging players in collecting scientific data and interviews in the areas they visit.

8.2.5 Geogames and Geoplay Can Help Narrow the Participation Gap

While there is a wealth of examples that demonstrate young people's ability to investigate and engage in the civic life of their local community, the path is not always easy, particularly within formal school settings. One challenge students can face when exploring and investigating particular aspects of their community is the existence of a *participation gap*. That is, young people don't always have the networks, expertise, or interest needed to conduct locally-focused investigations or participate in certain community-based actions (especially those associated with more formal civic institutions). In some instances, participation may require a longer time scale, the actions of many individuals, or access to specific forms of training or capital—all factors that can produce a barrier to participation or engagement. Not all of these challenges can be overcome in short order. For instance, in the case of civic planning, students in a classroom may become informed and involved, but only hold a small amount of power when it comes to directly impacting decisions. In many instances, community-based decisions require long-term time commitments, which students may not be able to meet. In part, this is amplified by the typical structure of schooling (e.g., time schedules that are not conducive to deep dives, especially ones tied to local studies and a reliance on courses that emphasize coverage of content over depth).

Games can help bridge this gap by providing a context or entry point for participation. They can allow players to inhabit new, otherwise inaccessible roles, from which new perspectives, possibilities, and structured challenges can be unlocked. Sometimes, this involves exploring local places through extreme roles, positions, and actions that would not be possible, or even recommended, in everyday life. Similarly, games can be used to simulate events and allow players to experience, explore, and shape imagined happenings. They can also amplify inputs, thereby allowing players to complete tasks that are beyond their level of expertise. Finally, games can be used to compress time scales. While an issue may play out in a community over several years or more, games can help us envision and even experience the consequences of our actions and decisions in a much shorter time frame.

Re-activism, developed by PET Lab, provides an example of how games can be used to scaffold players' participation in their community. Guided by a series of challenges and prompts, the game requires players to develop performances, visual works of art, or re-enactments at historical sites associated with important activist movements or events (Macklin and Guster 2012). The game can be easily modified and it has been run in a number of American cities, including Atlanta, Philadelphia, and New York. In instances of the game played in New York, for example, players visited the site of the Triangle Shirtwaist Factory fire, an important event in labor history, and the Stonewall Inn, an important site in the history of the gay rights movement. Instead of simply reading pre-existing plaques at these locations, players are encouraged to generate new stories and experiences. The "pop-up" forms of activism players create during the game disrupt the normal flow of places, and draw attention to the multi-layered histories and narratives associated with them.

In one way, the participation gaps mentioned here are realities to be acknowledged; limitations on what is possible with a place-based approach. But they can also be viewed as an invitation for reflection and change. How might the difficulty associated with enacting long-term interventions that engage students in official decision making processes serve as a criticism of existing structures and be leveraged to make change at a systems level? How might geogames allow students to "try on" leadership roles typically reserved for adults? Is this the type of participation we want to cultivate, or should we be focusing our efforts on exploring other forms of participation that fall outside of the traditional models of civic engagement? For example, in the next section, we suggest that engaging students in designing their own place-based games is itself a form of civic engagement.

8.2.6 Geogames, and the Design of Geogames, Can Foster Opportunities for Learners to Tell Their Own Stories of Place and Engage in Discussions and Actions That Impact the Future of Local Communities

While we have so far focused on games designed by researchers and professional media designers, when considering the potential of geogaming and its connections to place based learning, we feel it is equally, if not more important to develop resources and experiences that help others design their own games. Indeed, it is a natural transition for educators and students to move from playing geogames to *designing* geogames. The same tools that put digital game design within our reach also enable novices, or those without strong technical backgrounds, to make games. This is an important move because designing games typically requires deeper engagement with places and constructs, than simply playing games.

In our own work, emphasizing participant design has taken on many forms: from supporting the development of easy to use prototyping tools (Gagnon 2010; Holden 2015), to engaging our students in game design (Mathews 2010; Holden 2015), to collaborating with community organizations to co-develop games about issues and places that are important to them. Enacting and supporting geogame design activities requires action on many fronts and can involve students and learning in many different ways. For example, one of the key goals of the aforementioned **Up River** project was to spark educators and students to design their own games about the local watershed. In this model, we designed an initial experience, which took place in several unique local places, and then challenged students and teachers to build "expansion packs" for the game. In order to facilitate this process we held workshops where participants first played and critiqued **Up River** and then developed their own ideas for adding new quests, locations, characters, storylines, media, and scientific data to the original game. In the end, this approach helped us increase the scope of the original design and meet one of the main goals of the project; to develop people's understanding of the estuary and encourage them to share their own experiences, stories and perspectives.

Engaging students in designing games can align closely to many of the key goals of place-based education, especially when done using a student-centered approach. In addition to requiring students to investigate their community as part of the design process, building local games provides opportunities for them to share their own perspectives on local places and issues. Building games is also a natural form of interdisciplinary knowledge building and is an accessible activity that can be used to explore and represent authentic problems and engage new audiences in learning about their local community (Mathews 2010; Mathews and Holden 2012).

8.3 A Critical Perspective on the Design of Geogames

The games referenced in this chapter exemplify a specific type of geogame; one that reflects the values and goals of place-based education. They were developed and implemented with particular educational goals in mind, such as fostering critical awareness, inquiry and engagement. As a result, these "place conscious" games are different from geogames designed primarily for entertainment purposes, or games that simply layer data over places without considering local cultural values and practices. Indeed, the latter might actually run counter to a place-based approach, especially one grounded in a critical perspective.

Within this context, it is important for geogame designers to consider the potential impact of their games on players and the communities in which they are played. This is especially important when we, as designers and players, engage with places and cultures as outsiders. This does not mean that geogames should only celebrate local places. They can also be used to critically interrogate and challenge the status quo, for example, by exposing hidden or suppressed narratives, highlighting inequalities, and providing new avenues for civic participation. Still, as with other media, it is important to recognize the values and practices we promote through the design and play of our games. As part of this reflective process we should also:

- Be mindful of the fluidity and multiplicity of place;
- Thoughtfully consider how people, places, concepts, and issues are represented and remediated through the design and play of our games;

- Realize that a game is embedded in a larger culture, curriculum, etc.;
- Involve many stakeholders in the design and implementation of a game, especially people familiar with the places and themes it relates to; and
- Build and test in situ in order to raise awareness of possible conflicts and issues—i.e. don't build a game outside of the physical and social contexts it will be used in;
- Activate many kinds of learning through your design, not just recall of facts (e.g. cultivation of relationships, practical action, emergent exploration);
- Prepare players and debrief with them after the game.

8.3.1   If You Are Intrigued by These Ideas, How Can You Get Started?

While we see the potential of games and play to support place-based education, educators typically face many challenges when attempting to use them in their classrooms. These range from technological (e.g., dealing with rapid changes in digital technologies that can produce short obsolescence cycles), to institutional (e.g., fighting against curricular traditions that do not value or integrate the study of local places). Through our teaching, as well as our work helping others use and design geogames, we have encountered many of these challenges ourselves. Along the way, we have developed a few strategies that might be useful for newcomers interested in integrating place-based games into their own educational contexts.

- *Start small*—Create a simple game before embarking on a more complex design. Experimenting with many simple (and smaller) ideas or game mechanics helps build your fluency with the tools, deepens your understanding of the interplay between game elements and player experience, and sparks ideas for more dynamic work.
- *Build prototypes*—Test early and often. Iteratively improve designs by testing them with real people and in situ. Engage your students in testing and critiquing your designs.
- *Build on what you already have*—Existing activities or units can be amplified by incorporating a geo-gaming experience. Porting current activities to mobile also provides practice with using the tools and insight into what is lost and gained by bringing in "game-like" elements to an activity or moving from analog to digital modes.
- *Design for an event or place*—Develop a game around a pre-existing school or community event, environmental center, park, museum, neighborhood, etc. One of the nice things about designing for an event is that it means you have a pre-existing audience. This can assist you in extending your creations beyond the classroom.

- *Foster a culture of play*—Play can be a difficult to foster within formal situations. It cannot be enforced, only cultivated. One common form of play is roleplay, where players briefly take on new identities. Note that engaging in this requires an environment where learners feel safe taking risks and leaving existing identities behind.
- *Design around an issue, theme, concept, or question*—This is one of the most natural, simple, and authentic ways to bring local place and multiple perspectives into the picture. Do this instead of starting from and returning to subject-based topics.
- *Put away the digital tools*—Using pen-and-paper might not sound exciting, but they are accessible and flexible technologies. Simple, analog design also provides an opportunity for you and your students to explore the affordances of particular tools and mediums/media.
- *Experiment with freely available tools* (this part is play too)—Don't expect the perfect technological solution to be available. Whether considering software, hardware, or the articulation of a framework, try things that have promise and use them if they deliver, even if they don't give you everything you might want. Almost all of our games have included some mobile software, but have extended beyond the screen onto paper because the software available could not do everything we wanted.
- *Get involved with affinity spaces around these tools/uses*—Play and read about other people's games and borrow from their work. Find and talk to others using similar design tools. Share your own ideas and projects with the world.

It can also be instructive to *explore different configurations and trajectories*. There is not a single best way to build a game or series of games within a setting. Instead, we ask questions like: How does the game fit into the larger goals of the learning trajectory? What role does the game play in the larger learning ecology? How do the content and ideas present in the design, as well as the experiences players have, integrate with the larger area of inquiry they are involved in?

Games can provide a short encounter with a place or concept that can be linked to a larger idea or line of inquiry taking place back in the classroom. This can happen in multiple ways: A group of students, teachers, and community members engaging with a single neighborhood or local ecological area might play or design a series of mini-games, encouraging them to see and interact through a progression of spiraling or layered experiences. Or, students could play a single, extended game in one location, involving a series of quests or missions requiring them to explore the neighborhood across a number of visits, sometimes as a class and other times on their own as part of a homework assignment. The game could also begin in-class, before students go into the community, and involve both analog and digitally mediated interactions. Finally, a game could occur across multiple places, encouraging students to make comparisons, look for patterns, similarities, and differences.

8.3.2   How Can We Facilitate the Growth of These Ideas?

Above, we considered motivations for combining geogaming and place-based education, showed examples, and gave suggestions for newcomers. But, where does this path lead and what is necessary for these activities to gain traction? We believe an advantage of designing localized games is that they provide opportunities for educators and students to create mediated experiences tailored to local needs, interests, and contexts. This can provide an alternative to the production and distribution of curriculum materials by well financed publishers and distributors, or "delivered from above" by academic institutions. Indeed, producing locally-focused curricula and experiences that compliment or replace centralized production and distribution is a key method for supporting the tenets of place based education.

The real potential for geogames to support place-based learning, then, relies heavily on the ability for a wide range of people to produce locally focused designs. While there are many ways to support the development of localized games, we can broadly organize the overlapping types of growth needed to move forward into three areas: platforms, frameworks, and community.

Platforms More platforms are needed that provide non-programmers with opportunities to build their own rich, locationally interactive media. Would-be local game designers need authoring tools that are easy to learn, but also capable of supporting significant experimentation and depth. At least some of these should be designed for use across multiple contexts, not just for a specific setting or pre-defined literacy. Much like how advances in technology altered photography and video production, developing accessible authoring tools for easily producing geo-locative games has the potential to diversify who is able to create and what gets created.

Frameworks There needs to be continued work towards developing and researching frameworks and models for how to support people in designing their own games, especially those who have limited design experience. That is, simply providing tools does not solve the challenge of how to use them to produce quality games and related experiences. In part, this also highlights the need for further discussions about design processes and methods, game mechanics, characteristics of "good" games, and assessment of learning through design. Adopting place based education as a guiding pedagogy implies changes in epistemology and methodology, too. If our means of evaluation, or what we evaluate through place-conscious geogames does not diverge from assessing uniform subject matter (especially as a set of facts), we are missing the potential of this approach. In other words, what we try to measure should be closely aligned with the values that motivate our designs, and vice versa.

Community To increase diversity and bring new perspectives into this area, we need to broaden the range of people producing and playing geogames. We need wider representation across disciplines, including the arts, but also a stronger ethos of interdisciplinary and inclusive research and design. Building collaborative partnerships around these activities requires new relationships and configurations. If place-based geogames are going to provide a real alternative to the textbook model of learning, individual educators need to be able to participate in affinity spaces and collaborative design activities where their expertise and perspective is valued. If, on the other hand, geogames continue to primarily exist within the usual silos of educational research, and are only shared and discussed within academic contexts (e.g., journals and presentations), this work will have limited impact.

8.4   Conclusion

Place based education is not just a new subject to be added to the course catalog, but a set of values that articulates a non-traditional pedagogical stance. In some instances, when educators and designers use games to support place-based education, their work reifies traditional approaches to teaching and learning. In these instances, while geo-locative games are used, they are built on a model of information dissemination, thereby missing the inquiry-based and emergent nature of place-based learning. While there are times when a more didactic approach might be used within a place-based experience (and one we have certainly used in our own teaching), our hope is that our work, and that of the broader field, will more thoroughly take advantage of the affordances of locative, mobile, and game-based learning to promote student-centered and inquiry-based forms of teaching and learning. With that said, it is not our belief that this challenge can be met by simply building more and better games. Instead, larger questions and challenges associated with the current state of education (e.g., issues tied to local autonomy, funding, standardized testing) must be explored and addressed. It is, however, our hope that good game design and associated research, as well as inspiring stories from the field highlighting educators using geogames in robust ways, can inform these larger conversations about educational reform, especially the opportunities and challenges associated with place-based education. Getting to this point includes unpacking what has not worked and why; building accessible tools that invite locally-focused design; and cultivating communities of designers, practitioners, and researchers, at both the local and global levels who can support and amplify each other's work.

References

Dikkers S, Martin J, Coulter B (eds) (2012) Mobile media learning: amazing uses of mobile devices for learning. ETC, Pittsburgh

Gagnon D (2010) ARIS: an open source platform for developing mobile learning experiences. Master's thesis, University of Wisconsin

Gruenewald D (2003) The best of both worlds: a critical pedagogy of place. Educational Researcher 32(4):3–12

Gruenewald D, Smith G (eds) (2008) Place-based education in the global age: local diversity. Lawrence Erlbaum Associates, New York

Holden C, Sykes J (2012) Mentira: prototyping language-based locative gameplay. In: Dikkers S, Martin J, Coulter B (eds) Mobile media learning: amazing uses of mobile devices for teaching and learning. ETC, Pittsburgh, pp 113–129

Holden C, Dikkers S, Martin J, Litts B (eds) (2015) Mobile media learning: inspiration and innovation. ETC, Pittsburgh

Holden C (2015) ARIS: augmented reality for interactive storytelling. In: Holden C, Dikkers S, Martin J, Litts B (eds) Mobile media learning: inspiration and innovation. ETC, Pittsburgh, pp 67–83

Klopfer E, Squire K (2004) Getting your socks wet: augmented reality environmental science. In: Proceedings of the sixth international conference of the learning sciences, University of California, Los Angeles, 23–26 June 2004

Macklin C, Guster T (2012) Reactivism: serendipity in the streets. In: Dikkers S, Martin J, Coulter B (eds) Mobile media learning: amazing uses of mobile devices for teaching and learning. ETC, Pittsburgh, pp 151–169

Martin J (2009) Mystery trip. In: Dikkers S, Martin J, Coulter B (eds) Mobile media learning: amazing uses of mobile devices for teaching and learning. ETC, Pittsburgh, pp 99–110

Mathews J (2009) A window to the past: using augmented reality games to support historical inquiry. Paper presented at the American Educational Research Association annual meeting, San Diego, CA, 13–17 April 2009

Mathews J, Squire K (2009) Augmented reality gaming and game design as a new literacy practice. In: Tyner K (ed) Media literacy: new agendas in communication. Routledge, New York, pp 209–232

Mathews J (2010) Using a studio-based pedagogy to engage students in the design of mobile-based media. English teaching: practice and critique 9(1):87–102

Mathews J, Holden R (2012) Place-based design for civic participation. In: Dikkers S, Martin J, Coulter B (eds) Mobile media learning: amazing uses of mobile devices for teaching and learning. ETC, Pittsburgh, pp 151–169

Mobile UW (n.d.-a) Field research project: WeBird. https://mobile.wisc.edu/mli-projects/field-research-project-webird/. Accessed 3 Sept 2014

Mobile UW (n.d.-b) Project: SustainableU. https://mobile.wisc.edu/mli-projects/project-sustainable-u/. Accessed 3 Sept 2014

Orr D (1994) Earth in mind: on education, environment, and the human prospect. Island Press, Washington, DC

Rosenkrantz H (2014) Jewish time travel gets real. The Covenant Foundation. http://www.covenantfn.org/news/152/Jewish-Time-Travel-Gets-Real. Accessed 3 Sep 2014

Smith G, Sobel D (2010) Place and community based education in schools. Routledge, London

Squire K, Dikkers S (2012) Amplifications of learning: use of mobile media devices among youth. Convergence 18(4):445–464

Squire K, Jan M (2007) Mad city mystery: developing scientific argumentation skills with a place-based augmented reality game on handheld computers. J Sci Educ Technol 16(1):5–29

Squire K, Jan M, Mathews J, Wagler M, Devane B, Holden C (2007) Wherever you go, there you are: place-based augmented reality games for learning. In: Sheldon B, Wiley D (eds) The design and use of simulation computer games in education. Sense Publishing, Rotterdam, pp 265–296

Wagler, M. & Mathews, J. (2012). Up River: Place, ethnography, and design in the St. Louis River Estuary. In S. Dikkers, J. Martin, & B. Coulter (Eds.). Mobile media learning: Amazing uses of mobile devices for teaching and learning (41–60). Pittsburgh, PA: ETC Press

Chapter 9
A Cost-effective Workflow for Depicting Landscapes in Immersive Virtual Environments

Nathaniel J. Henry

9.1 Introduction

Geogames draw inspiration from two technologies: geographic information science (GIS) and video games. These technologies share a common history that stretches back to the invention of ancient "map games" such as chess and Go. Both fields have advanced rapidly in the past two decades, spurred by the rapid expansion of computer graphical and processing capabilities (Ahlqvist 2011). GIS analysts and video game designers might be surprised by the similarity of their data management techniques: both rely on hierarchical data management structures, employ techniques for minimizing processor loads when representing complex scenes, and use layers to organize their data sources (Shepherd and Bleasdale-Shepherd 2009). Overlapping engagements with simulation, multi-party collaboration, and the web indicate that GIS and video game technologies may be headed down converging paths (Ahlqvist 2011).

Despite these commonalities, GIS and video games diverge sharply in their representations of space and place. Shepherd and Bleasdale-Shepherd (2009) explain this difference in terms of reality, the extent to which a technology describes the world as it actually exists, and realism, which describes the representational system used to display game objects and phenomena. King and Krzywinska (2003) add another factor for consideration: the perspective through which end users view and interact with the represented environment. The user could view the world through the eyes of a virtual character (first-person perspective) or peek over that character's shoulder (third-person perspective). Otherwise, they might observe the world at a distance by looking down from an oblique or vertical angle. GIS and video game scholars have labeled the final two perspectives as "managerial" or "god's-eye" views, and one need only play top-down games such as **Sim City** or

N.J. Henry (✉)
The Ohio State University, Columbus, OH, USA
e-mail: henry.557@osu.edu

© Springer International Publishing Switzerland 2018
O. Ahlqvist, C. Schlieder (eds.), *Geogames and Geoplay*, Advances in Geographic Information Science, https://doi.org/10.1007/978-3-319-22774-0_9

Civilization to understand why: in these games, the user often exerts greater control over the game landscape, and keeps a constant watch for decades or centuries of in-game time.

When compared on the axes of reality, realism, and perspective, video games exhibit a greater variety in representational forms and content than GIS. Video games can depict worlds ranging from semi-historical to entirely fantastic; their depictions can be as symbolic as geometric shapes or almost as naturalistic as the real world; and players can experience games from any of the perspectives listed above and more. On the other hand, technical and thematic constraints have historically limited GIS technology to representing the world as it exists, using largely symbolic forms of representation such as markers and color gradients, from a top-down or oblique perspective (Shepherd and Bleasdale-Shepherd 2009). Additionally, realistically depicting real-world landscapes may require prohibitively high levels of effort and technological specialty for most GIS research groups. Geogames may have adopted video game technologies and formats, but current implementations often follow the reality-realism-perspective combination dictated by conventional GIS.

In gaming, representation and perspective are more than just virtual window dressing; they ultimately determine how players will experience all phenomena in the game world. Representational style and perspective factor into the player's sense of presence, defined as "a physical sensation of complete submersion in a digital medium" (Denisova and Cairns 2015). In turn, the player's sense of presence in the game-world can facilitate immersion, a feeling of engrossment with the narrative content of the game. Empirical studies confirm the relationship between realism, representation, presence, and immersion: according to studies conducted Denisova and Cairns (2015), people were more immersed in gameplay when they experienced it from a first-person perspective, regardless of stated preference. Additional research from Sylaiou et al. (2010) found a statistically significant correlation between a user's sense of presence in a virtual reality environment and their enjoyment of the scenes being depicted. Certainly, surface realism and viewpoint alone are insufficient to spark immersion. Other key contributors to game immersion include intuitive game interfaces, narrative, and character-building (Taylor 2002; Bayliss 2007). However, games rendered in a naturalistic style and viewed from a first-person perspective may facilitate more engaging narrative types and stronger connections with in-game characters, further increasing the likelihood of player immersion.

This chapter outlines an inexpensive workflow for representing real-world landscapes as three-dimensional (3D) virtual environments in a video game engine. By relying on inexpensive methods of data collection and an automated process for 3D reconstruction based on computer vision software, the workflow lowers the barriers for high-fidelity video game representations of existing landscapes. The following section defines key terminology relating to three-dimensional game environments. It then describes existing representations of real-world places, spaces, and phenomena in virtual environments, and explores commons advantages and disadvantages of these implementations. Next, the workflow is described in detail. Finally, the workflow's implications for geogame creation are discussed and possible extensions are proposed.

9.2   Modeling the World in Three Dimensions: Immersive Virtual Environments and Their Applications

Geogames are not the first applications to model real-world landscapes in a three-dimensional virtual environment; a number of researchers have already attempted to do so on behalf of commercial, academic, and government organizations. Understanding the successes and challenges of these attempts can inform the ways that geogames engage with 3D technologies.

While the following examples might differ in format, content, and purpose, they all aim to represent existing landscapes and phenomena in immersive virtual environments. A virtual environment can be broadly defined as a computer-generated spatial environment (Magee 2011). In the context of this chapter, immersive virtual environments refer to virtual environments formatted in a way that facilitates user immersion: that is, virtual environments rendered in three dimensions, tending towards naturalistic rather than symbolic representations, and allowing the user to view the world from a first-person or third-person perspective.

Virtual environments encompass a broad range of possible forms and content types, so specific implementations are often better described with a more specific sub-category. Virtual worlds, for example, are multi-user virtual environments which emphasize social interaction. In virtual worlds such as **Eve Online**, **Second Life**, and **Active Worlds**, users can simulate real or fictional lives and interact with others through the use of in-world avatars (Loke 2015). On the other end of the spectrum, serious games and simulations impart contextual knowledge or skills on users by engaging them with representations of specific real-world situations (Cain and Piascik 2015; Šimic 2012). And while most users currently engage with 3D digital landscapes through a screen, the fast-developing field of virtual reality aims to provide more intuitive tools for experiencing and manipulating virtual environments (Magee 2011).

Commercial video games have been representing real-world landscapes in three dimensions since the late 1990s, when the iconic video game protagonist Lara Croft navigated the bank of the River Thames and the roof of St. Paul's Cathedral as part of the classic **Tomb Raider** series. However, this chapter will focus on more recent experimental applications of 3D virtual environments in academia, health services, and the military. In these fields, researchers have created virtual representations of the real world in order to solve a problem or improve an existing process, from recruit training to historic preservation. These implementations can be categorized into two types: the educational, archaeological, and archival applications, which tend to focus on user engagement and learning; and the military, health care, and emergency response applications, which tend to focus on user training and professionalization. Implementations of the same category tend to encounter similar types of successes and challenges due to their shared purposes and related institutional contexts. Given the tendency for past geogames to focus on learning and exploration, three-dimensional geogames would likely fall into the first of these categories.

9.2.1   Virtual Landscapes as a Tool for Engagement and Learning

For educators, archaeologists, and historians, three-dimensional virtual environments hold promise as a pedagogical tool. According to Dickey (2005), if education is a social practice, then video games and virtual worlds can serve as social platforms, enabling activities and narratives that lead to learning. Users in virtual environments can interact with types of data and knowledge representations that are not simply not possible in a classroom setting. Additionally, virtual learning environments may be a more attractive learning option for a younger generation of "digital natives" (Chau et al. 2013). With these goals in mind, a number of researchers have taught courses set in online virtual worlds, where students attend lectures in reconstructed classrooms, "meet up" with their avatars to complete group assignments, and even explore the virtual equivalents of their college campus together (Dickey 2005; Ritzema and Harris 2008; Fominykh et al. 2011; Chau et al. 2013).

Archaeologists and historians in particular have engaged with 3D technology as a new way to share and explain the past. Virtual environments and virtual reality hold great potential as storytelling tools: like a good story, they can transport the user into a world conceived by the "author" and impart meaning, resulting in greater empathy and understanding of the content being displayed. There is also a hope that virtual environments will attract a wider audience of people, especially young people, to explore their own history (Dawson et al. 2011). Past applications of 3D within the fields of archaeology and archival studies have included a digital reconstruction of the Parthenon in Athens, a digital museum of Chinese culture in anticipation of the 2008 Beijing Olympics, and the depiction of Thule whalebone houses and Siglit-Inuvialit sod houses in virtual reality (Punzalan 2014; Pan et al. 2009; Dawson et al. 2011).

Several educational studies concluded that courses in virtual worlds can enhance aspects of student engagement and enjoyment. Dickey (2005) describes how one of her courses surged in popularity after a virtual world version was created. Chau et al. (2013) found that students who participated in a virtual world course reported higher levels of satisfaction than their offline counterparts, and were particularly happy with the flexibility and geographic freedom offered by the online course. In terms of emotional engagement, Dawson et al. (2011) also received strong positive feedback about their virtual reality reconstructions of whalebone and sod houses. After a group of Inuit Elders viewed the reconstructed dwellings in 3D, they reported an increased sense of connectedness with their past. As one commented, "all the stories I used to hear when I was young are coming back to me. It really makes me think about what it would have been like to live in my ancestors' home." (Dawson et al. 2011).

However, if a user experiences technical issues, it can transform the 3D experience into a frustrating endeavor far worse than traditional methods of learning. Chau et al. (2013) identified connectivity issues and difficulties learning the user interface as two common pitfalls for students enrolled in their virtual course. Students with less technical savvy face the greatest barriers to engagement in virtual world-based courses, and they can easily fall behind if the instructor does not intervene. Finally, in situations where fidelity to the real world is important, dishonest replication could lead to digital "forgeries" that misinform users about the topic at hand (Punzalan 2014).

9.2.2 Virtual Environments as a Tool for Training and Professionalization

Serious games that depict real landscapes and situations have already received significant attention as potential training platforms for medical, military, and emergency response professionals. Serious games and simulations have an obvious appeal when it comes to training exercises: as researchers have noted, "the use of simulation is a safe and inexpensive way to prepare and educate people on how to respond to emergencies" (Sharma and Otunba 2012). Using serious games and simulations, these fields can acclimate players to stressful situations, test costly equipment, and reproduce complex systems that would be impractical or dangerous to emulate in real life (Roman and Brown 2008; Boosman and Szczerba 2010). Past 3D applications have included simulators for airplane evacuation, flooding on a naval ship, unit navigation in an urban environment, and networked medical devices in a hospital ward (Sharma and Otunba 2012; Hussain et al. 2009; van der Hulst et al. 2013). Within NATO alone, militaries from at least five countries have adopted serious games as a core training platform (van der Hulst et al. 2013). Military strategists are also attempting to implement tools from the developing field of virtual reality in order to gain a competitive advantage over their adversaries (Magee 2011).

Evidence indicates that virtual environments can serve as an effective vehicle for contextual training and learning. A number of studies have concluded that serious games can successfully impart knowledge and attitudes on military recruits (discussed in Roman and Brown 2008). Naval recruits who participated in a 3D flooding control simulation displayed more confidence during a real-life test than a control group who attended a class on the topic (Hussain et al. 2009). On the other hand, learners and trainees may not always receive the intended lesson: clunky artificial intelligence, lack of proper physics modeling, and the realism of rendered objects may render some serious games and simulations useless for medical, military, or emergency response training (van der Hulst et al. 2013). Additionally, on-screen simulations do not always replicate the stress of actual emergency situations, so some learners may not take these virtual lessons seriously (Sharma and Otunba 2012: 572).

9.2.3 Cost and Feasibility Challenges to Digitally Recreating the Real World

When it comes to the cost and feasibility of creating and implementing immersive virtual environments, training applications in the medical, military, and emergency response fields are subject to different considerations than applications intended for education or community outreach. Despite their high cost, 3D training applications offer a cheaper and safer alternative to testing expensive equipment or recreating potentially dangerous scenarios in person. As a result, their expected cost is actually lower than the current systems in place (Sharma and Otunba 2012). Large organizations such as medical institutions or militaries possess the computational resources and technical capabilities to generate large areas of realistic terrain and simulate complex processes (Roman and Brown 2008; Boosman and Szczerba 2010). The major barriers to 3D virtual training applications relate to institutional acceptance and the current lack of standard procedures for creating virtual environments (Magee 2011).

Educators, archaeologists, and historians face the opposite set of problems: fewer institutional barriers stand in the way, but virtual environments are often too costly or technically difficult for widespread implementation. Educators who currently wish to construct their campus in a virtual environment must choose between expending an enormous amount of time building it by hand, or else create an automated reconstruction using expensive equipment such as a laser scanner. Tools for visualizing virtual environments, such as 3D headsets and CAVE devices, can be expensive and are not available to most educators (Loke 2015). For programs that operate on limited budgets, Dawson et al. (2011) worry that the time and money used to construct digital representations of the world would be better spent on programs that provide material content and benefits to target groups.

Three-dimensional geogames would likely experience successes and face challenges similar to those found in existing educational, archaeological, and archival implementations. By representing landscapes as immersive virtual environments, geogames could reap the benefits in user enjoyment, presence, and learning outcomes reported by these implementations. However, if geogame designers are forced to reconstruct real 3D landscapes from scratch, they will likely encounter the same intractable feasibility issues.

As an alternative to these methods, the next section introduces an automated workflow for digitizing existing landscapes into immersive virtual environments. Instead of using expensive or technically difficult tools, this process relies on simple data collection technologies and a user-friendly computer vision software. It takes far less time to complete than manually reconstructing a scene in a virtual environment, and the equipment involved is less expensive than standard 3D scanning equipment. Taken as a whole, this workflow is expected to reduce the barriers for representing high-fidelity virtual environments in geogames.

9.3   Using Kite Aerial Photography and Computer Vision Software to Create Immersive Digital Environments

This process draws on multiple existing workflows for kite aerial photography and 3D scene reconstruction. Geert Verhoeven (2009, 2011) describes a number of methods used to obtain low altitude aerial photographs, and then details how these aerial photographs can be reconstructed into a three-dimensional digital surface using computer vision software. Olson et al. (2013) employ this process to record an archaeological site at a high spatial and temporal resolution, with an emphasis on quickly processing and storing three-dimensional data. Using a similar workflow, Currier (2015) creates a high-resolution orthographic photo mosaic of a remote Indonesian island, then assesses the resolution and accuracy of photo mosaics and digital elevation models (DEMs) that can be produced using this process. Aspects from each of these studies will be discussed in greater detail below. This chapter extends existing methods by importing a photo mosaic and DEM into a free, widely-used video game engine. The result of this workflow (Fig. 9.1) is a landscape that can be used as the basis for a playable game level.

The workflow is outlined below, and more specific technical considerations are enumerated within Table 9.1.

During the summer of 2014, this process was employed to digitize a stretch of beach-side cliffs at Campus Point in Santa Barbara, California. The data collection, scene reconstruction, and playable landscape creation will be discussed in turn, using the digitization of the Campus Point cliffs as an example.

9.3.1   Data Collection

Verhoeven (2009) describes the many data collection options available for capturing low-altitude aerial photographs of a study area. While booms and poles, balloons, kites, and unmanned aerial vehicles (UAVs) can all be employed, the most appealing option for many researchers is also one of the oldest: a camera rig attached to a kite. Kite aerial photography (KAP) dates back to the late 1800s; today, it is a highly popular method for aerial photography due to its portability, durability, and low cost (Verhoeven 2009). Unlike balloon photography, kite aerial photography requires only a single initial purchase; and in most areas of the United States, kites can be flown up to 150 m above the ground without regulations, while UAVs are subject to a set of stricter and frequently-changing regulations (Currier 2015). Additionally, for data collection in populous areas, negative perceptions of UAVs may color public opinion regarding the research, whereas kites are far more likely to spark positive interest and engagement. This survey used a delta kite with a width of 3.4 m, along with two tails for stability in high winds.

Kites do introduce limitations into the circumstances under which data can be collected: the survey area must be fairly open and free of aerial obstructions such as telephone wires or tree branches. Additionally, kites can only be operated with a steady wind speed of 10–25 km/h. In this sense, they are somewhat complementary to drones and balloons, which require low wind speeds for successful operation. However, given these two conditions, a kite can successfully and reliably lift the camera rig used for aerial photo capture.

Attached to the kite is a rig holding a downward-facing camera that shoots pictures at regular intervals. The rig itself is a simple metal frame that can be readjusted to

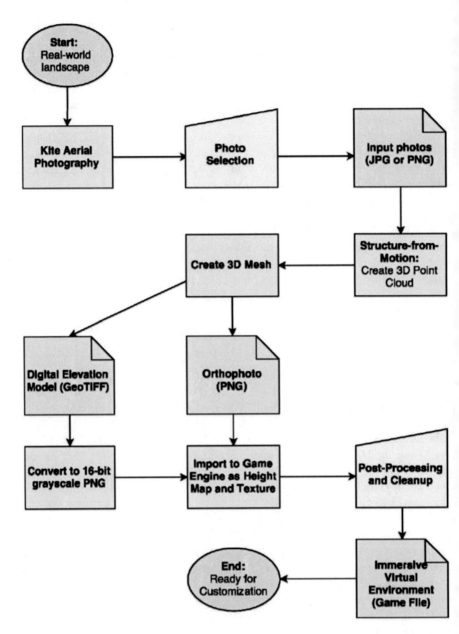

Fig. 9.1 A general workflow for depicting a real-world landscape in an immersive digital environment

angle the camera in various directions. The camera is attached to the metal rig using a screw, and the rig is attached to the kite with a picavet cross, a suspension system designed to maximize the stability of the rig. The camera itself is an inexpensive

Table 9.1 Equipment, best practices, and settings used within the workflow

Processing step	Hardware and software	Specifications used
Kite aerial photography	11-ft Delta Kite; Canon Powershot S100 with the Canon Hack Development Kit; KAP Rig	*Camera*: Focal distance of infinity; shutter speed of 1/500 s; f/8 aperture; ISO greater than 400 (Smith et al. 2009, in Currier 2015) *Control Points*: Set at least three; optimal results with ten or more (Agisoft 2016) *Recording:* KAP rig hung approximately 20 m below the kite (Currier 2015); pictures taken at 10 s intervals from both vertical and oblique angles
Photo selection	Photo viewing program	Photos should have 80% overlap on a path, 60% side-by-side overlap; photos with blurry or moving objects should be excluded (Agisoft 2016)
Creating point cloud	Agisoft PhotoScan	*Aligning Cameras*: High accuracy; "Generic" preselection *Dense Cloud*: Medium quality (to speed up processing)
Creating 3D mesh	Agisoft PhotoScan	*Mesh*: Height Field surface type; Extrapolated interpolation; Dense Cloud as source data. *Texture:* Orthophoto mode; Mosaic blending *DEM Output*: GeoTIFF with square power-of-two dimensions (such as 4096 × 4096 px) *Orthophoto Output*: PNG with same dimensions as the DEM
DEM conversion	FIJI	Convert the DEM GeoTIFF to a 16-bit grayscale PNG image
Importing to game engine	Unreal Engine 4	*DEM*: Import the PNG image using the Landscape "import from file" option (Epic Games 2015a) *Orthophoto*: Import by creating a new Material with a "tile size" equal to the Landscape dimensions, then applying it to the Landscape as a base color (Epic Games 2015b)
Post-processing and cleanup	Unreal Engine 4	Rotate and resize the landscape to scale using the object properties; erase areas with no data using the landscape editing toolbar (Epic Games 2015c); adjust key game controls such as the level's starting point

Canon point-and-shoot device running the Canon Hack Development Kit, a firmware package that extends the functionality of Canon cameras (CHDK User Manual 2016); using this kit, an interval timer can be installed on the camera at no cost.

The Campus Point cliffs are in many ways an ideal location for kite aerial photography, due to the regular sunny weather, consistent wind, and lack of aerial obstructions. Additionally, because most of the terrain is composed of rock, sand, or dense low-lying vegetation, the cliffs are also a suitable subject for computer vision reconstruction, which can be confused by shifting objects (Agisoft 2016). However, although the game engine only imports square elevation models, the terrain could only be recorded as a long, thin strip; while the cliffs run for kilometers from the University of California at Santa Barbara to the nearby city of Isla Vista, they are bordered by the ocean on one side and a lagoon on the other, allowing for less than 30 m of navigable space in some areas. This meant that a square DEM was exported with many areas of no elevation data, and these unmapped areas were later removed from the video game landscape. Ultimately, a stretch of cliffs approximately 500 m long and 50 m across was processed for this study.

Aerial photographs were collected by repeatedly flying the KAP rig, at an elevation of approximately 100 m, across the study area in a straight line. Each pass across the study area was parallel and overlapping with previous passes. Additionally, to better capture rough terrain, the camera angle was adjusted to collect downwards-facing and oblique shots on different passes (Verhoeven 2011). Over the course of several runs, more than 1000 aerial photographs of the study area were collected.

In addition to aerial photographs, a number of ground control points were collected using brightly-colored markers and a GPS device. The markers were created using high-contrast colors in order to simplify identification from a distance. For this study, five ground control points were collected using both temporary and permanent markers. Latitude, longitude, and elevation were collected for each marker using a GPS device.

9.3.2 Three-Dimensional Reconstruction

After collecting aerial photographs and ground control points, the study area was reconstructed as a three-dimensional point cloud, and then as a mesh, using computer vision software. Put simply, computer vision is a science that recreates three-dimensional objects from two-dimensional images using mathematical algorithms, a process sometimes referred to as "structure from motion" (Verhoeven 2011). According to Olson et al. (2013), much like the human brain can comprehend a three-dimensional object by viewing it from multiple angles, an algorithm can determine relative positions of points in 3D space by viewing them across multiple images. A software employing computer vision can use these spatial relationships to construct a set of points in 3D space, known as a point cloud. After an initial, "sparse" point cloud is constructed, the software uses it as a reference to reconstruct a more detailed set of geometries known as the "dense" point cloud (Olson et al. 2013). Finally, the software joins the dense point cloud together into a mesh and overlays textures and colors retrieved from the input images, recreating a 3D surface that is geometrically and visually similar to the original object (Agisoft 2016).

This study used Agisoft PhotoScan Pro, a computer vision software that employs both photogrammetric and computer vision algorithms to reconstruct 3D surfaces. While many other options exist for recreating 3D objects, from free software such as 123D Catch and VisualSFM to the more expensive PhotoModeler Scanner and 3DM Analyst Pro, Agisoft PhotoScan Pro offers several advantages for 3D geogame creation. PhotoScan Pro sports an intuitive user interface that handles all aspects of the 3D reconstruction process, from adding photos to exporting the model. More importantly, the software allows users to georeference and export Digital Elevation

Models, a crucial aspect of this workflow. The software has been well-documented in past studies and workflows (including Verhoeven 2011; Gatewing 2012; Olson et al. 2013; Currier 2015), allowing for further instruction and optimization.

Of the aerial photographs taken over campus point, 200 were selected for 3D reconstruction. These photographs did not include blurry or moving objects, represented an appropriate amount of overlap between images, and included both oblique and downward-facing shots of the study area. The photographs were loaded into the software, where they can be viewed and edited. Several photographs showing the sky, moving objects such as people, or irregularly-shaped objects such as trees that would be poorly represented by a digital elevation model were masked and excluded from processing. The result of photo alignment is a sparse point cloud, the skeleton of the final model (Fig. 9.2). The sparse point cloud of the Campus Point cliffs contained 19,952 points before cleaning.

After the photos have been aligned, ground control points can be marked and georeferenced in the original photographs. While each ground control point must be located in at least two input images to be correctly referenced, PhotoScan will automatically place markers in other images based on the photo alignment. After the markers have been placed and assigned Longitude, Latitude, and height, and expected accuracy, root mean square errors for the ground control points can be calculated. The user can also remove erroneous points in the sparse point cloud to improve the model's final geometry, and use the bounding box to restrict areas for further processing (Agisoft 2016).

Once the model has been georeferenced and cleaned, the user can automatically create the dense point cloud using the sparse point cloud for reference. Since this step creates three-dimensional points based on all matched pairs between photos, the dense point cloud can end up having millions of vertices: the dense cloud for the Campus Point cliffs contained over 65 million points. This point cloud bears a striking resemblance to the original imagery even at a close range (Fig. 9.3).

After the dense point cloud has been created, the user must carefully align the bounding box with the region of the study area that is to be exported into the game engine. The bounding box must be horizontal, with the plane of projection (represented by the red edge of the box) below the landscape. After properly aligning the bounding box and manually removing any erroneous points, the user can generate a high-resolution mesh, and then cover that mesh with a textured mosaic of the aerial photos (Fig. 9.4).

Computer vision is resource-intensive; a model of several hundred photos can take days to process from start to finish. Several steps can be taken to reduce processing time: after aligning photos, users can split the project into "chunks", each of which can be processed faster separately than as part of the overall model. Olson et al. (2013) recommend splitting projects with over 1000 photographs into chunks to reduce processing load. Users can also configure a dedicated GPU for 3D reconstruction, accelerating the process (Agisoft 2016). Additionally, users can choose to set the quality parameters at each step of the reconstruction to lower values, as slight variations in output quality may be negligible compared the increase in processing time (Olson et al. 2013).

Fig. 9.2 The sparse point cloud generated by Agisoft PhotoScan's structure-from-motion algorithm. The *blue rectangles* show the estimated position and alignment of the camera for each input photograph

9.3.3 Exporting to a Game Engine

Once a 3D model has been created, it can be exported as a wide variety of 3D formats, from .OBJ to .FBX and even .PDF. However, for compatibility with a game engine, two derived 2D images must be produced instead: a Digital Elevation Model (DEM) and an orthorectified photomosaic of the study area, also called an orthophoto. These two outputs can be converted into video game elements and deployed in Unreal Engine 4, an industry-leading game engine. This section will discuss Unreal Engine 4, the PhotoScan image exports, and how a few minor file conversions can allow for interoperability between the two software packages.

Game engines are software applications that act as frameworks for game development. They render low-level processes such as object collision, loading, and user input, allowing for game developers to focus on the actual narrative and experience of a game (Ward 2008). With their reusable chunks of code and more accessible designer interfaces, game engines can help new developers quickly get a working game off of the ground. They are also invaluable assets for 3D game designers thanks to their automatic rendering of physics and 3D navigation. Using a game engine, geogames designers can quickly and easily turn 3D models into playable game levels.

Fig. 9.3 The dense point cloud generated in Agisoft PhotoScan

Fig. 9.4 The dense point cloud is connected to become a three-dimensional surface

This study renders landscapes in Unreal Engine 4, the latest in a line of widely-used Unreal game engines. While other game engines could plausibly display this landscape, Unreal Engine 4 has several features that may appeal to geogames designers. The engine is free for most noncommercial research and education applications; its learning curve may be slightly friendlier, thanks to a large developer network and a scripting interface written in C++ (Masters 2015); and the game engine provides native support for virtual reality headsets such as the Oculus Rift, opening up further possibilities for immersive virtual reality in geogames.

Unreal Engine 4 requires two types of inputs to import a landscape: a height map, which describes the landscape's contours, and materials, which describe the colors and textures of that landscape. First, the DEM must be exported from the 3D model and converted into a format compatible with the game engine. This application used FIJI, an open-source image editing software, to convert the image to the proper format (Schindelin 2012). Finally, the converted image was imported directly into Unreal Engine 4 as a Landscape surface (Epic Games 2015a), which can be resized, rotated, and edited using in-game tools (Epic Games 2015b, c).

After the landscape has been appropriately formatted, it must be covered with textures from the 3D model in order to approximate appearance of the study area. This is accomplished by exporting an orthorectified photograph, or orthophoto, from the 3D model. Orthorectification is "a technique in which a photograph is differentially corrected using a DEM" (Currier 2015). Because of the correction, the resulting image or image mosaic corresponds exactly with the contours of the associated DEM. This photo can be imported into Unreal Engine 4 as a material, which can simply be dragged and dropped onto the map to recreate the subject area in high detail. Using this process, a 4096 × 4096 px orthophoto export from PhotoScan was converted into a material and placed on the landscape in Unreal Engine 4 (Fig. 9.5). Despite the oblong shape of the study area, which limited the possible dimensions of the orthophoto and DEM, the resulting ground features had a resolution equivalent to 6 cm.

By placing the "level start" action somewhere on the landscape, the user is able to navigate a three-dimensional representation of the study area in the first person. In Fig. 9.6, the landscape is shown from the perspective of the game avatar. The resolution of the height map and texture are high enough for the user to recognize individual landscape features.

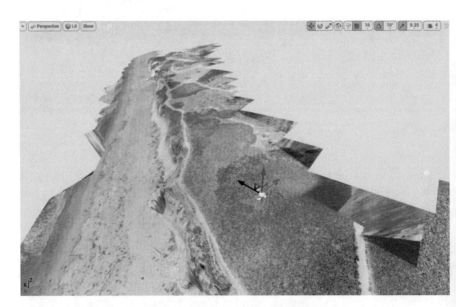

Fig. 9.5 The Campus Point cliffs recreated in the Unreal Engine 4 game engine

Fig. 9.6 From a first-person perspective, the user can identify ground objects (*left*) and larger terrain features such as a rocky outcropping (*right*)

9.4   Discussion

This chapter has already discussed the potential impact of 3D immersive geogames and explored a workflow for reconstructing existing landscapes in three dimensions using computer vision software and a game engine. However, the benefits and pitfalls of this specific workflow have not yet been explored. Is this method worth the trouble of mastering kite aerial photography, photogrammetric reconstruction, and a new method for game design?

From the perspectives of cost, time, and user experience, I argue that this workflow has much to offer geogames. All the major expenses of this process, including a kite, KAP rig, camera, GPS, and peripheral accessories can be obtained for less than US $1000 in total. These costs are incurred only once, and they provide the geogames creator with the tools to digitize new areas indefinitely. In terms of time spent, once the designer has mastered all aspects of data collection and transformation, the entire process should take less than 2 days of work. The majority of that time would be spent on kite aerial photography, which is pleasant, and processing in PhotoScan, which can be automated. Reductions in time and money spent lower the barriers for nonprofit and educational groups who desire to create immersive 3D content but are concerned about the costs involved.

In terms of user experience, the workflow described in this chapter allows users to reconstruct existing landscapes at a resolution that is high enough to recognize individual ground features. The user navigates this landscape in the first person, and Unreal Engine 4 allows developers to set intuitive controls for the end users. Because this workflow only creates the template upon which a game can be built, geogame developers still must construct engaging narratives and gameplay if they hope to bring about user immersion. However, at the very least, this workflow allows end users to navigate and experience game environments in a way that may not have been possible from a top-down perspective.

Several technical issues still limit the sizes and types of landscapes that can currently be represented using this workflow. Because of processing constraints, Unreal Engine 4 cannot efficiently process landscapes with a height map or material resolution greater than 8192×8192 (Epic Games 2015c). This means that landscapes of a certain size cannot be digitized at a resolution high enough to accurately represent landscape features. Due to the downwards-facing orientation of orthophotos, steep vertical surfaces in the landscape are poorly textured and may be hard to clearly recognize. Additionally, as mentioned before, vegetation is not adequately represented by kite aerial photography alone (Olson et al. 2013: 259). Many of these problems could be greatly improved by representing landscape objects, including trees, rocks, and even cliffs, as discrete objects lying on top of a lower-resolution landscape. If these features were processed as Unreal static mesh objects, a data type that is more interoperable with 3D formats, then aerial photographs could be combined with shots from ground level to create game objects with a far higher resolution. By improving ground object recognition, geogames can move one step closer to accurate representation of the real world.

Throughout the workflow, I found myself balancing trade-offs that pitted mesh quality against processing speed, landscape extent against visual resolution, and terrain variety against in-game comprehension. The game engine's editing tools also presented me with a choice between manually enhancing the terrain and maintaining fidelity to the original remote sensing data. Navigating these trade-offs ultimately required me to make judgment calls based on the perceived preferences of an end-user. As geogames mature, developers will increasingly encounter these types of subjective challenges when making choices about visual representation, user experience, and gameplay; future research should embrace these challenges as a chance to develop the more artistic and experiential aspects of geogames.

Because game engines are designed with customization in mind, this workflow also opens the door to a number of possible extensions. Games can be published for PC, iOS, Android, and the web, allowing users to browse and experience virtual environments through a variety of interfaces. Geogame levels are already being tested in conjunction with virtual reality; trial tests with the Oculus Rift headset have elicited extremely positive feedback from participants (Fig. 9.7). Game engines also facilitate the addition of sound and haptic (touch) feedback to a level, which could

Fig. 9.7 A student navigates the landscape using a virtual reality headset (*left*); stereoscopic images sent to the headset create the sensation of three-dimensional navigation (*right*)

expand the senses that players use to navigate the virtual environment. To further reduce the barriers for nonprofit and educational game creators, a similar workflow using only free software tools should be developed. Finally, the three-dimensional format opens up new possibilities for in-game storytelling and interactions; these narrative possibilities should be explored and linked to user immersion.

References

Agisoft LLC (2016) Agisoft PhotoScan user manual: professional edition, Version 1.2. http://www.agisoft.com/pdf/PhotoScan-pro_1_2_en.pdf. Accessed 1 Feb 2016

Ahlqvist O (2011) Converging themes in cartography and computer games. Cartogr Geogr Inform Sci 38:278–285. https://doi.org/10.1559/15230406382278

Bayliss P (2007) Beings in the game-world: characters, avatars, and players. In: Proceedings of the 4th Australasian conference on interactive entertainment. RMIT University, pp 1–6

Boosman F, Szczerba RJ (2010) Simulated clinical environments and virtual system-of-systems engineering for health care. In: Interservice/industry training, simulation, and education conference. pp 1–9

Cain J, Piascik P (2015) Are serious games a good strategy for pharmacy education? Am J Pharm Educ 79:1–6

Chau M, Sung W, Lai S, Wang M, Wong A, Chan KWY, Li TMH (2013) Evaluating students' perception of a three-dimensional virtual world learning environment. Knowl Manag E-Learn 5:323–333

CHDK User Manual (2016) Canon Hack development kit community. http://chdk.wikia.com/wiki/CHDK_User_Manual. Accessed 1 Feb 2015

Currier K (2015) Mapping with strings attached: kite aerial photography of Durai Island, Anambas Islands, Indonesia. J Maps 11:589–597. https://doi.org/10.1080/17445647.2014.925839

Dawson P, Levy R, Lyons N (2011) "Breaking the fourth wall": 3D virtual worlds as tools for knowledge repatriation in archaeology. J Soc Archaeol 11:387–402. https://doi.org/10.1177/1469605311417064

Denisova A, Cairns P (2015) First person vs. third person perspective in digital games: do player preferences affect immersion? In: Proceedings of the ACM Chicago 2015 conference on human factors in computing systems. ACM, New York, pp 145–148

Dickey MD (2005) Three-dimensional virtual worlds and distance learning: two case studies of Active Worlds as a medium for distance education. Br J Educ Technol 36:439–451. https://doi.org/10.1111/j.1467-8535.2005.00477.x

Epic Games, Inc (2015a) Creating and using custom heightmaps and layers. In: Unreal engine 4 documentation. https://docs.unrealengine.com/latest/INT/Engine/Landscape/Custom/index.html. Accessed 6 Aug 2015

Epic Games, Inc (2015b) Editing landscapes. In: Unreal engine 4 documentation. https://docs.unrealengine.com/latest/INT/Engine/Landscape/Editing/index.html. Accessed 6 Aug 2015

Epic Games, Inc (2015c) Texture support and settings. In: Unreal engine 4 documentation. https://docs.unrealengine.com/latest/INT/Engine/Content/Types/Textures/SupportAndSettings/index.html. Accessed 6 Aug 2015

Fominykh M, Prasolova-Forland E, Morozov M, Gerasimov A (2011) Virtual campus in the context of an educational virtual city. Int J Interact Learn Res 22:299–328

Gatewing (2012) Software workflow: AgiSoft PhotoScan Pro 0.9.0. http://uas.trimble.com/sites/default/files/downloads/gatewing_PhotoScanworkflow090.pdf. Accessed 6 Aug 2015

van der Hulst A, Muller T, Besselink S, Vink N (2013) The potential of serious games for training of urban operations. In: STO modeling and simulation group conference. Sydney, Australia, pp 1–8

Hussain TS, Roberts B, Menaker ES, Coleman SL, Pounds K, Bowers C, Cannon-Bowers JA, Murphy C, Koenig A, Wainess R, Lee J (2009) Designing and developing effective training games for the US Navy. In: Interservice/industry training, simulation, and education conference. pp 1–17

King G, Krzywinska T (2003) Gamescapes: exploration and virtual presence in game-worlds. In: Digital games research conference proceedings. p 109

Loke S (2015) How do virtual world experiences bring about learning? A critical review of theories. Australasian J Educ Technol 31:112–122

Magee L (2011) Virtual Reality (VR) as a disruptive technology. In: Defence Research & Development Canada Defence Research reports. http://cradpdf.drdc-rddc.gc.ca/PDFS/unc121/p536948_A1b.pdf. Accessed 6 Aug 2015

Masters M (2015) Unity, source 2, unreal engine 4, or CryENGINE which game engine should i choose? Digital tutors. http://blog.digitaltutors.com/unity-udk-cryengine-game-engine-choose/. Accessed 6 Aug 2015

Olson BR, Placchetti RA, Quartermaine J, Killebrew AE (2013) The Tel Akko Total Archaeology Project (Akko, Israel): assessing the suitability of multi-scale 3D field recording in archaeology. J Field Archaeol 38:244–262. https://doi.org/10.1179/0093469013Z.00000000056

Pan Z, Chen W, Zhang M, Liu J, Wu G (2009) Virtual reality in the digital Olympic museum. IEEE Comput Graph Appl 29:91–95

Punzalan RL (2014) Understanding virtual reunification. Libr Q Inform Commun Policy 84:294–323

Ritzema T, Harris B (2008) The use of second life for distance education. J Comput Sci Coll 23:110–116

Roman PA, Brown D (2008) Games—just how serious are they? In: Interservice/industry training, simulation, and education conference. pp 1–11

Schindelin J (2012) Fiji: an open-source platform for biological-image analysis. Nat Methods 9:671–675. http://fiji.sc/. Accessed 6 Aug 2015

Sharma S, Otunba S (2012) Collaborative virtual environment to study aircraft evacuation for training and education. In: 2012 International conference on Collaboration Technologies and Systems (CTS). pp 569–574

Shepherd IDH, Bleasdale-Shepherd ID (2009) Videogames: the new GIS? In: Lin H, Batty M (eds) Virtual geographic environments. Esri, Beijing, pp 311–344

Šimic G (2012) Constructive simulation as a collaborative learning tool in education and training of crisis staff. Interdiscipl J Inform Knowl Manag 7:221–236

Smith MJ, Chandler J, Rose J (2009) High spatial resolution data acquisition for the geosciences: kite aerial photography. Earth Surf Process Land 34:155–161. https://doi.org/10.1002/esp.1702

Sylaiou S, Mania K, Karoulis A, White M (2010) Exploring the relationship between presence and enjoyment in a virtual museum. Int J Hum Comput Stud 68:243–253. https://doi.org/10.1016/j.ijhcs.2009.11.002

Taylor LN (2002) Video games: perspective, point-of-view, and immersion. Dissertation, University of Florida

Verhoeven G (2009) Providing an archaeological bird's-eye view an overall picture of ground-based means to execute low-altitude aerial photography (LAAP) in archaeology. Archaeol Prospect 16:233–249. https://doi.org/10.1002/arp.354

Verhoeven G (2011) Taking computer vision aloft—archaeological three-dimensional reconstructions from aerial photographs with PhotoScan. Archaeol Prospect 18:67–73. https://doi.org/10.1002/arp.399

Ward, Jeff (2008) What is a game engine? Game career guide. http://www.gamecareerguide.com/features/529/what_is_a_game_.php. Accessed 6 Aug 2015

Chapter 10
Structural Gamification of a University GIS Course

Michael N. DeMers

10.1 Introduction

College level Geographic Information Systems (GIS) courses are, like most university courses, linear, rigid, punitive, and self-contained. Most have laboratories involving the use of high-end, complex GIS software with a steep learning curve. This chapter demonstrates one way to convert one such laboratory course into a quest-based learning (QBL) environment. The chapter illustrates how I am currently using **3DGamelab**'s Quest-Based Learning Management System (LMS) to convert this traditional, college course into one that provides choice, rewards for learning from mistakes, and opportunities for credential building. I describe how to incorporate labs normally requiring face-to-face interaction, how to provide choice and still cover the material, how to leverage outside learning material, and how to encourage life-long learning. Many of the rewards, especially the badges, are linked explicitly to the US Department of Labor's geospatial technology industry's competency model, thus encouraging learners to see this as an opportunity to plan for future employment. Finally, this chapter discusses some of the difficulties of course mapping a complex, full-semester course. In particular it illustrates issues of adapting to learner needs, and the time and effort required to construct and deliver an interactive-intensive learning environment. I make recommendations for solutions and adaptations for such difficulties.

Increasingly instructors of all university courses, including Geographic Information Systems courses are experiencing pressure to change to online formats of course delivery. Unfortunately the common perception still remains that by merely placing the content traditionally delivered face-to-face in a nicely ordered structure in the Learning Management System for the learner to retrieve is sufficient

M.N. DeMers (✉)
New Mexico State University, Las Cruces, NM, USA
e-mail: demers01@gmail.com

© Springer International Publishing Switzerland 2018
O. Ahlqvist, C. Schlieder (eds.), *Geogames and Geoplay*, Advances in
Geographic Information Science, https://doi.org/10.1007/978-3-319-22774-0_10

for such courses to be effective. Fortunately research regarding the design of online courses has matured greatly and has resulted in a set of standards for course design that is specifically intended to improve online course design. Called the QM, or Quality Matters Rubric,[1] it is a national peer review and certification process that allows course designers to receive feedback on the following eight general standards of course design:

1. Course overview and introduction
2. Learning objectives,
3. Assessment and measurement
4. Instructional materials,
5. Course activities and learner interaction,
6. Course technology,
7. Learner support and accessibility
8. Usability

Alignment with such standards improves the likelihood of creating an environment in which effective online learning can occur. What it does not guarantee, however, is that the course design structure will, in and of itself, motivate the learner to engage in the material. Despite effective course QM Certification and the direct-employment implications of practical courses like GIS, students do not always find the material particularly engaging. They do not always find the pace appropriate or sufficient learning choices available. Perhaps most importantly they seldom if ever experience a grading structure that rewards hard work and learning, but does not punish the learner for making mistakes as long as they can demonstrate that they have learned from them. Additionally, most GIS courses focus heavily on the hard technical skills of the geospatial industry while downplaying or ignoring the necessary soft skills, especially those revolving around communication (DeMers 2012). The search for a way to combine the rigor of a traditional lab-based GIS course with the allure, freedom, choice, and incentive-rich game environment suggests that the content of a traditional GIS could be ported to a quest-based environment that would provide these characteristics and lead to enhanced engagement in learning (Friedemann et al. 2015).

10.2 Literature Review

QBL is part of a larger set of approaches to learning called game-based learning. It is important to note immediately that while games may be included in game-based courses there is a fundamental difference between the use of games as learning tools and the general process of gamification (Mallon 2013). Gamification refers to the adoption of some or all of the typical mechanics of games that carry with them the allure and addictive behavior (Banfield and Wilkerson 2014; Renaud and Wagoner

---

[1] The Quality Matters Higher Education Rubric can be found at https://www.qualitymatters.org/rubric.

2011; Rouse 2013). The mechanics of concern are points (experience points), badges, levels, leaderboards, challenges and other incentives and reward structures that motivate gamers.

While all of the mechanics are present to a greater or lesser degree how these mechanics are implemented is highly dependent on the form of gamification in play. There are two forms of gamification—content gamification in which the course content is converted to games and structural gamification in which the content remains intact and the mechanics are modified to leverage the same incentives found in games. In content gamification exercises are converted from their tradition formats—e.g. reading assignments, writing assignments, lectures, discussions, labs, etc.—to actual games in which the object is to win the game, and as a by-product, learn what the instructor wants. The games can be action/adventure games in which the concepts and skills being taught are necessary components of the game. They can be simulations of real world disciplinary settings in which the learner is involved in role-play. Strategy games also find themselves being used in the educational environment in which the strategy itself is composed of the content being taught. Whatever type of game is being employed, the focus of content-gamification is to set up a scenario in which the learner is not always aware of the intended learning objectives, but is rather being tricked into learning the material (DeMers 2005, 2010).

Structural gamification does not convert the content into games but rather focuses on the game mechanics that are considered some of the more common reasons that games are so inherently addictive. While generally, but not exclusively, not focusing on making the learning itself "fun," instead it focuses on the manner in which the course is organized. Such structural gamification means that all assignments, while traditional in their methods of delivery, are considered to be quests to be conquered. While they often have prerequisite skills and knowledge, acquired through success-ful completion of other quests, there is far more flexibility as to when they can be taken. In effect the course, if based on a text, does not require that the learner neces-sarily move linearly through the material. Within the loose structure the learners have choices regarding taking high value, long assignments versus lower value but much shorter timeframes. Such an approach allows a college level instructor the ability to provide at least some of the important characteristics of the personal learn-ing environment (PLE), particularly the self-regulation (Dabbagh and Kitsantas 2012) that is increasingly being demanded of today's millennial learner (Dede 2007).

Virtually all the remaining mechanics are related to grading. Unlike normal assignments, quests are all-or-nothing in that if the learner achieves a certain level of accomplishment (normally considered 85%) the quest is considered successful. This is identical to the testing procedures of the Esri online tutorial modules (Johnson and Boyd 2007). One crucial pedagogical improvement of this approach to grading is that each time a quest is returned, the learner receives feedback from the instructor regardless of the success or failure of the quest. This greatly enhances the amount of student-faculty interaction—a feature considered highly correlated to student success (Lamport 1993; Kuh and Hu 2001).

Collectively, there is also a fundamental difference in how course grades are accumulated. In a typical course one has a limited number of points and as the

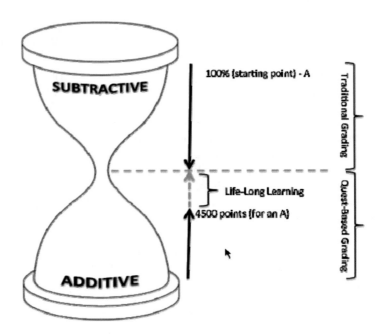

Fig. 10.1 The difference between the subtractive grading process of traditional courses versus the additive grading process of Quest-Based courses

course progresses the learner continues to lose points—moving from a beginning score of 100% toward an ever decreasing score. In quest-based learning the process is reversed in that there are generally more points (called experience points in game parlance) than one needs to achieve a grade of A in the course. In short, QBL style grading is additive while traditional grading is subtractive (Fig. 10.1). Other grade-related components include a reward structure that encourages not just completing quests, but doing exceptional work, working hard to complete difficult tasks, etc. Each of these "rewards" is built into the system as incentives for desired behaviors. Some of these rewards are "achievements" based on completing a substantial portion of the course material. Other rewards are badges that document micro credentials for particular skills, behaviors, characteristics, or knowledge that is considered useful based on industry needs related to the course material. Badging has been shown to be useful as incentives for self-efficacy, enjoyment, and motivation in education (Ahn et al. 2014; Chen et al. 2012; Denny 2013; Finkelstein et al. 2013; Gibson et al. 2013). One added advantage of badging is that organizational websites such as Mozilla Backpack provide a means to store, retrieve, and most importantly to share these badges beyond the classroom setting for documenting selected micro-credentials with peers, colleagues, current employers, or future employers.

Competition is also encouraged by allowing students to compete with each other for the accumulation of quest experience points, levels, and awards. In video games, one's scores are compared to those who are also engaged in the same game but not necessarily as adversaries. Such comparisons, called "leaderboards" provide a direct comparison of ones success in the game to others. In some games and in most educational settings the leader board is an opt-out item for those who do not wish to have their names included. At the same time, such anonymous members are still able to view others' names on the leaderboard to gauge their level of success.

Given the nature of millennial learners (Dede 2005) so commonly found in university undergraduate courses and their constant exposure to social media and computer games it seems reasonable to assume that such technology would be rapidly embraced by educators in general and college educators in particular but this does not seem to be the case (Dicheva et al. 2015). While business, marketing, corporate management and even wellness industries have adapted to gamification methods, educational adoption seems to be more of an emerging trend. Johnson et al. (2014), however, are hopeful that these technologies will soon be on the "adoption horizon" as they call it within 2–3 years.

The missing piece to this adoption puzzle is a game-based platform; a learning management that incorporates the mechanics of structural gamification. There are few mature systems capable of being able to allow gamification mechanics, and even fewer that are specifically designed around such an approach. One of the most accessible mature systems designed around the quest mechanics of quest based learning is **3DGamelab**, a product of a company called Rezzly. The **3DGamelab** is a product of years of research in game-based learning dating back to efforts to use virtual worlds as a platform for game-based learning (Dawley 2011; Dawley and Dede 2012). With time the efforts of the researchers at Boise State University developed the principles, constructs, and mechanics of an operational quest-based learning management system (Haskell 2012, 2013). Because this platform is so new, there is little if any formal documentation regarding the success or failure of implementing courses using this approach, although there is one blog that discusses the approach briefly (Kolb 2015). This chapter is a first step in filling that intellectual void with particular reference to Geographic Information Systems education.

10.3 Mechanics

As discussed earlier there are a few learning management systems capable of supporting the structural gamification needed to implement a truly quest-based course. Among the most mature, most robust, and most fully developed is the **3DGamelab** produced by Go Go Labs (now called Rezzly.com). Beyond its level of development, I selected this platform because it is based both on experience and educational research (Dawley 2011) and because its price per student was reasonable for this experiment.

The **3DGamelab** is a quest-based LMS that is not just compatible with structural gamification but is based entirely on it. It provides a badging system, rewards, leveling up, leaderboards, and the mechanics to allow these features to be operationalized in a course setting. These allow the course developer to concentrate on the content rather than on these mechanics. The **3DGamelab** is integrated into existing LMS software, in this case Canvas, so the learner doesn't have to move between elements of Canvas and **3DGamelab**.

10.3.1   Design Considerations

Because quest-based learning provides a high level of choice both in what learning opportunities the students pursue and in the sequencing, the orchestration of the many moving parts of a complete laboratory course becomes a bit of a challenge. Fortunately, the **3DGamelab** user interface provides a roadmap for designing your QBL course. To map out the quests, **3DGamelab** the roadmap—an Xcel-based template that calculate experience points, tabulates sequencing and prerequisites, estimates time requirements, and provides other feedback for course development. It also calculates total quest points, available reward points, total possible points, winning condition (number of points for a grade of A), total number of quests, and the estimated time to complete all the quests (in minutes) (Fig. 10.2).

As with any course there is a need to determine how much weight to assign to both types of learning activity (e.g. lecture material versus laboratory or project) as well as to each activity. I was guided as much by experience as anything but the general approach was based on the idea that one should be able to approach, but not obtain a grade of A from doing the basic lectures, labs, portfolios, quizzes, and exams. The remaining of the points require the learner to select up to 400 points from a total of 3700 points in quests that remained available. This provides plenty of material for additional learning even when the semester is concluded. As the course is often the specific numbers will be adjusted based not on the amount of time that I estimated each quest to consume, but rather by tabulating averages for each quest as the system compiles them.

As the instructor creates each quest it is included in the table and properties related to whether the quest requires grading by the instructor or is graded automatically, the estimated time for each quest to be completed, number of experi-

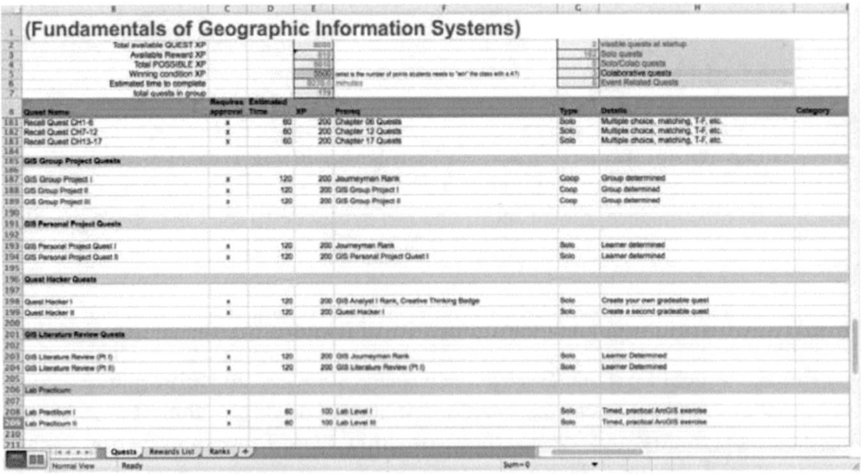

Fig. 10.2 A portion of an Xcel-based QBL roadmap with individual quest information in the table rows and the summary data at the *top*

ence points, and the prerequisites are included and both the time and experience points are updated in the summary portion of the table. To further assist with course navigation I employed a branching diagram, not unlike a mind map, to keep track of course prerequisites and learning pathways. Some QBL instructors use a software package for tablets called Popplet®. I use a similar package called Scapple® that works on PC's or Macs. The choice of software is a matter of familiarity, hardware, and personal preference. One reason I use the Scapple® software is that it allows me to use the icons I use in the course to identify the quests rather than relying on text that, for a course of this magnitude, would create a diagram that would be far too difficult to read.

10.3.2 Quests

The system allows the user to create individual quests that can carry up to 200 experience points depending on the anticipated length of time or amount of effort required to complete the assignment. For this course I created the following quests:

- 4 introductory quests designed to teach the use of the system
- 74 lecture quests (17 book chapters broken into smaller topical sections)
- 27 laboratory quests (based on a commercial laboratory book)
- 11 laboratory portfolio quests (based on the laboratory book tutorials)
- 1 portfolio compilation quest (collecting the artifacts into a course eportfolio).
- 20 Esri tutorial quests
- 17 recall quests (chapter quizzes)
- 3 unit recall quests (mid-term exams)
- 3 group project quests (parts 1, 2, and 3)
- 2 personal project quests (parts 1 and 2)
- 2 quest hacker quests (make your own quest)
- 2 literature review quests (parts 1 and 2)
- 2 laboratory practicum quests
- 2 course feedback quests (to provide me with ways to improve the course)

One important aspect of the system is that it allows for not only creating the quests (assignments) but also a mechanism for learners to submit their work for evaluation within the quest-based grading model (Fig. 10.3). Because this is a quest-based course the assignments are not graded as you would a class exercise. Instead they use the 85% rule in which the instructor determines a whether or not the assignment has achieved 85% of the learning objectives of the quest. If they have, the learner's quest is approved, usually with additional helpful comments. Alternatively, if the instructor feels that the assignment does not meet the 85% rule the assignment is returned with comments, helpful suggestions, and even additional sources of information to assist learning. This process continues until the learner completes the quest. Completed quest receive 100% of the experience points for

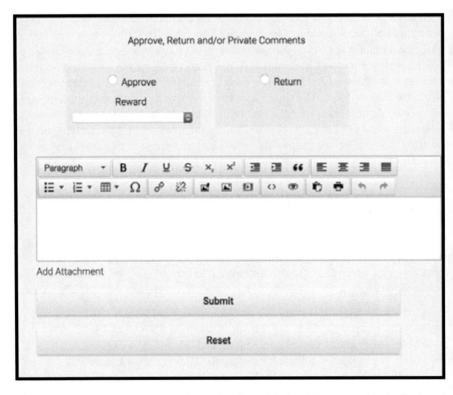

Fig. 10.3 3DGamelab Quest approval user interface with the ability to provide feedback and attachments if needed

each. This is as true of quizzes (recall quests) as of any other quest, in that the learner is allowed to take quizzes as many times as they need to pass.

The course content comprises traditional content variously emphasizing a large proportion of the GIS&T Body of Knowledge (DeMers 2009). The rewards for this course are grouped into three general categories—*awards*, generally related to lecture and laboratory completion as well as special accomplishments such as excellence, speed, hard work, acquisition of GIScience level of understanding and others; *achievements* (levels); and *badges* (Fig. 10.4 , DiBiase, et al. 2010). The badges are particularly of interest to this course because each badge is directly linked to one of the 25 competencies of the US Department of Labor Geospatial Skills Competency model. These badges that can be archived in the Mozilla badging backpack, together with the e-portfolio quests and the Esri online course certificates add incentives for the students by providing them with a vast array or credentials the students can bring to potential employers. It was hoped that the implementation such a quest-based GIS approach to learning would encourage greater participation, more interaction, more self-efficacy among the students, and ultimately more enjoyment.

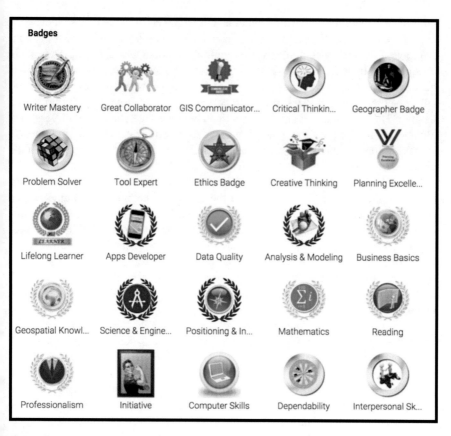

Fig. 10.4 Twenty-five badges linked to the 25 skills in the US Department of Labor Geospatial Skills Competency Model

10.4   Results

Developing a course of this scope, with nearly 180 interconnected quests is a complex process requiring a substantial commitment of time. Converting the narrated lectures into interactive multi-media content was by far the most time-consuming task. The laboratories however were surprisingly straightforward as the Esri laboratory manual used had the exercises fully laid out and included learning objectives and requisite products. Prior to each set of laboratory quests was a related tutorial that, in past courses, the students often skipped, making them struggle with the labs and costing the teaching assistants an inordinate amount of additional tutoring. The portfolio exercises were designed to correct this issue by requiring the student to turn in the products of the tutorials. The results were generally successful although many students did not immediately embrace the idea, and, only after learning how to navigate the system did they appreciate the tutorial.

Because the course material is less structured than a traditional course, as expected by experiences with this same online courses prior to gamification, a bimodal distribution of student engagement emerged early in the course. For week 3, n = 24 participants, the range of quest scores was 15 for a low and 1115 for a high. The mean for experience points was 332 and the median was 210. These numbers are paralleled by a high of 28 quests and a low of 2 quests. The students self-report the actual time they spend questing. One student has not reported these numbers so with n − 23 the average quest time is reported as ranging from 88 to 2 min with a mean of 31.5 min and a median of 30 min. There are nine students with less than 100 quest points and these same students started questing late; often waiting nearly 2 weeks before beginning a single course experience. Those heavily engaged in the content and those less than enthusiastically engaged.

By week 10, 6 of the 23 students had dropped the class—notably within ±5% of the rate of the previous five offerings of the course online but without quest-based learning mechanics. Two students had achieved GIS Analyst (modeler) rank, requiring 3400 experience points (Table 10.1). Notably there seemed to be no pattern of Badge achievement, but achievements exemplified by completing higher level quests or performing exceptionally on assignments were higher among the higher ranked learners. The latter trend is to be expected as higher achieving students will also be those who not only do assignment, but do them very well and achieve more quests resulting in additional achievement-based points.

When the course concluded 12 of the students had achieved the rank of Chief Executive Officers (over 4500 experience points), a rank associated with a grade of A in the course. Four of the students achieved a rank of GIS Master (between 4000 and 4500 experience points), a rank associated with a grade of B. Finally, one student did not complete the course, receiving a grade of incomplete. This latter student requested an extension of the course due to illness and is, at the time of this writing, completing the work not completed during the semester. There were no C, D, or F grades in the course, consistent with the findings of Haskell (2013), but since the conclusion of the course none of the students have continued to quest nor to pursue badging microcredentials. This is at odds with Haskell's results and suggests differences resulting from student level, quest engagement, circumstance, or course design. These results have yet to be investigated and remain important questions in the deployment of QBL courses.

Each quest allows students to rate it and to provide some feedback regarding any aspect of it they wish. Generally the preliminary students rated the quests with average scores of approximately 4.2 out of 5 and few provided verbal input. Those few who did seemed concerned about the amount of time required to perform each quest. In the case of the laboratory quests this seems a reasonable reaction but the students seems particularly concerned that the lecture exercises required too much time for the points achieved. In informal conversations with students regarding this, their concern was not just the amount of time required, but that they feared they would not be able to achieve the necessary 5500 experience points to achieve an A, or 4000 to achieve a grade of D. One objective of the course was to require the students to spend more time on the content as there is a direct positive correlation between time on task and learning. However, based on student feedback I reduced the point scores for each level and for final grades by 1000. This value seemed more reasonable and the learners began participating with more vigor, indicating that the high levels needed for grades acted as a disincentive for participation.

Table 10.1 Results of quest activity in week 10 of a 15-week GIS course

GamerTag	Rank/grade	TotalXP	QuestXP	Active	Completed	Dropped	RewardXP	Badges	Achievements	Awards
mastergeomex	GISAnalystII(modeler)	3495	3145	1	83	0	370	0	8	12
4gingkos	GISAnalystII(modeler)	3405	3035	5	81	1	390	5	9	10
desert bat	GISDatabaseManager	3055	2515	3	63	1	560	1	7	13
evitolas	GISAnalyst1	2835	2475	7	69	2	360	6	6	8
rebecca_armenta	GISAnalyst1	2755	2455	16	70	1	300	1	6	10
mrmusta	GISJourneyman	2605	2315	6	68	7	290	2	5	10
nirome	GISMasterTechnician	2145	1945	3	54	6	200	1	4	7
kevinvirden	GISTechnician	2095	1815	2	50	1	310	0	4	9
marcusal	MapReader	1485	1395	1	38	1	90	0	2	4
dil2.0	MapReader	1405	1035	6	29	1	370	11	5	5
jacobvallee	MapReader	1355	1255	3	38	5	100	0	2	5
tlowe	MapReader	1325	1255	5	39	0	70	1	2	2
carmertsandiego	MapReader	1315	1245	19	38	5	70	0	2	3
dsrtrosy	MapReader	1245	1095	7	31	1	150	2	2	5
imjustaguy	MapReader	1205	1095	9	34	0	130	1	2	4
mvela	MapReader	1205	1075	6	32	1	130	2	2	4
snoopylupe	MapReader	1175	1055	5	30	2	120	1	2	6

As the students progressed they eventually became more efficient at the lecture assignments that were primarily designed to get them to answer the questions at the end of the chapter in their own words. Initially I found that as they began taking the recall quests the number of students who passed them with a score of 9/10 (the closest a ten question quiz can be to the magical 85%) on the first try seemed considerably higher than students who took the traditional course. This indicates at least anecdotally that there is more learning happening in the quest course than the traditional course. As the weeks have gone on I have also noticed that the student responses to the lecture quest questions are becoming more robust, more detailed, and generally more insightful. I also noticed that I was able to award far more "awards of excellence" as a result of this.

A major issue of the course appeared early on and is continuing. Some learners purposely procrastinate; using the loose structure of the course to focus on other coursework. While this was part of the design it indicates that some constraints on quest completion needs to be applied. The teaching assistants are particularly adamant about imposing some restrictions on when material must be turned in, as they are responsible for grading the laboratory exercises. The procrastination required them to be familiar with all of the 27 laboratories—an unnecessary burden on the teaching assistants. An anticipated side effect of the deliberate procrastination is that those who are regularly handing in exercises will likely be forced to wait for grading feedback as large groups of exercises from the latecomers pile up.

At the beginning of the course the general consensus of the students was that they liked the idea of flexible scheduling. Unfortunately as the course began they also showed a lack of motivation to pursue the material aggressively to complete the course early, opting instead to wait. One positive feedback from informal discussions with the students was that they were surprised at the amount of feedback they received. This indicates that this part of the course implementation was at least recognized if not effective at enhancing the students' learning.

10.5 Summary and Conclusions

This course demonstrated few basic results that are notable, some positive, some negative. It appears that there is enhanced learning as evidenced both by the steadily increasing quality of assignments turned in by the students and the improved efficiency in the recall quests. As the semester progressed the answers to questions related to course content were increasingly more robust, more complete, and in the case of the better students, more insightful. This indicates a deeper level of comprehension of the material than what I have seen in previous, non-QBL incarnations of the course. The interactions between faculty and student were times greater than in both the face-to-face versions of the course as well as the previous online versions which content required less personal attention of the instructor to grade.

Procrastination was a major problem especially for the laboratory assistants who are responsible for grading the labs. While the feedback was constant, the amount of grading was substantially higher than one would expect of a normal college course. This has both negative and positive consequences. Obviously grading nearly 2000 quests requires a dedication to grading on the part of the instructor, however the constant feedback provides an astonishing amount of faculty to student interactions—interactions often touted by face-to-face instructors as the reason for not wanting to move to online instruction. Overall the amount of time spent providing feedback seems to have paid off in much enhanced learner comprehension and this is after all the primary reason for education.

Leveling seemed to be a motivating factor as students commented individually that they appreciated knowing what point scores they needed to achieve to obtain a specific grade. For some students the achievement of badges had meaning but these tended to focus more on the acquisition of Esri certificates of completion than on the other badges. Unfortunately, the allure of the Esri certificates resulted in many students spending an inordinate amount of time on these quests which, by design, provided far fewer points per time investment. The purpose of assigning such low scores was to discourage such an approach, as the acquisition of ArcGIS expertise is given a much lower value in the course over conceptual learning. As this activity was observed, intervention in the form of sending the learners a reminder of the need to concentrate on high value learning experiences over low value if they hoped to achieve desired grades. While the course roadmap had been provided to the learners at the beginning of the course, it was necessary to reintroduce the roadmap as well as to provide specific examples of how to achieve major rewards for less effort. This was especially important as I pointed out that the course contained Easter eggs (hidden rewards in this case) that provided huge rewards (e.g. 100 points) for achieving select benchmarks—e.g. completing an exam representing 1/3 of the lecture content.

Overall the course seems to have a qualified success with some obvious issues needing to be addressed. Beyond the procrastination issue, some learners find the nature of the exercises, especially the lecture exercises requiring them to provide responses in their own words, to be tedious to the point of being painful. This might suggest the use of audio responses to minimize the amount of typing on the part of the learner. This approach has its own negative consequences as it increases the amount of time it takes for grading. As the course continues more feedback will be available. The students will also be asked to participate in a filmed, unstructured exit interview to provide additional insights.

References

Ahn J, Pellicone A, Butler BS (2014) Open badges for education: what are the implications at the intersection of open systems and badging? Res Learn Technol 22:23563. https://doi.org/10.3402/rlt.v22.23563

Finkelstein J, Knight E, Manning S (2013) The potential and value of using digital badges for adult learners: DRAFT for public comment. American Institute for Research, Washington, DC

Banfield J, Wilkerson B (2014) Increasing student intrinsic motivation and self-efficacy through gamification pedagogy. Contemp Issues Educ Res 7(4):291–298

Chen Z-H, Liao CCY, Cheng HNH, Yeh CYC, Chan T-W (2012) Influence of game quests on pupils' enjoyment and goal-pursuing in math learning. Educ Technol Soc 15(2):317–327

Dabbagh N, Kitsantas A (2012) Personal learning environments, social media, and self-regulated learning: a natural formula for connecting formal and informal learning. Internet High Educ 15(1):3–8

Dawley L (2011) Questing across the spectrum of virtuality: aggregating learning in 3D GameLab. Int J Gaming Comput-Mediat Simul 3(4):iii–iiv

Dawley L, Dede C (2012) Situated learning in virtual worlds and immersive simulations. In: Spector JM, Merrill DM, Elen J, Bishop MJ (eds) Handbook of research on education communications and technology, 4th edn. Springer, New York

Dede C (2005) Planning for neomillennial learning styles. Educ Q 28(1):7–12. https://net.educause.edu/apps/eq/eqm05/eqm0511.asp?print=yes. Last accessed 30 March 2014

Dede, C. (2007). Reinventing the Role of Information and Communications Technologies in Education. In L. Smolin, K. Lawless, & N. Burbules (Eds.), Information and Communication Technologies: Considerations of Current Practice for Teachers and Teacher Educators [NSSE Yearbook 2007 (106:2), pp. 11–38. Malden, MA: Blackwell Publishing.

DeMers MN (2012) Preliminary analysis of GIS workforce education: linking practitioners with educators. Papers Appl Geogr Conf 35:41–49

DeMers MN (2010) Coyote teaching in geographic education. J Geogr 109(2):1–8

DeMers MN (2009) Using intended learning objectives to assess curriculum materials: the UCGIS body of knowledge. J Geogr High Educ 33(Supp 1):S70–S77

DeMers MN (2005) Coyote medicine: tricking your GIS students into learning. Proceedings ESRI education user conference, San Diego. http://proceedings.esri.com/library/userconf/educ05/papers/pap1934.pdf

Denny P (2013) The effect of virtual achievements on student engagement. Proceedings, SIGHI conference on human factors in computing systems, CHI13. pp 763–772

DiBiase D, Corbin T, Fox T, Francis J, Green K, Jackson J, Jefress G, Jones B, Brent J, Mennis J, Schuckman K, Smith C, Van Sickle J (2010) The new geospatial technology competency model: bringing workforce needs into focus. Urisa J 22(2):55–72

Dicheva D, Dichev C, Agre G, Angelova G (2015) Gamification in education: a systematic mapping study. Educ Technol Soc 18(3):75–88

Friedemann S, Baumbach L, Jantke KP (2015) Textbook gamification transforming exercises into playful quests by using webble technology. Proceedings, 7th international conference on computer supported education. http://www.researchgate.net/publication/278672297

Gibson D, Ostashewski N, Flintoff K, Grant S, Knight E (2013) Digital badges in education, 2015. Educ Inf Technol 20:403–410

Haskell CH (2012) Design variables of attraction in quest-based learning. PhD dissertation, Boise State University. http://scholarworks.boisestate.edu/cgi/viewcontent.cgi?article=1286&context=td

Haskell CH (2013) Understanding quest-based learning. [Whitepaper]. https://works.bepress.com/chris_haskell/21/. Retrieved 27 May 2016

Johnson AB, Boyd JM (2007) Content, community, and collaboration at ESRI virtual campus: a GIS Company' perspective on creating an online learning resource. J Geogr Higher Educ 29(1):115–121

Johnson L, Adams Becker S, Estrada V, Freeman A (2014) Games and gamification. In NMC horizon report: 2014 higher education edition. The NewMedia Consortium, Austin, TX, pp 41–43

Kolb L (2015) Epic fail or Win? Gamifying learning in my classroom, Eductopia blog. http://wwwedutopiaorg/blog/epic-fail-win-gamifying-learning-liz-kolb. Last accessed 7 Sept 2015

Kuh G, Hu S (2001) The effect of student-faculty interaction in the 1990s. Rev High Educ 24(3):309–332

Lamport M (1993) Student-faculty informal interaction and the effect on college student outcomes: a review of the literature. Adolescence 28(122):971–990

Mallon M (2013) Gaming and gamification. Public Serv Q 9(3):210–221

Renaud C, Wagoner B (2011) The gamification of learning. Princ Leadersh 12(1):56–59

Rouse KE (2013) Gamification in science education: the relationship of educational games to motivation and achievement. PhD dissertation, University of Southern Mississippi

Chapter 11
Geocaching on the Moon

Cheng Zhang

Abbreviations

AR	Augmented Reality
ARTEMIS	Acceleration, Reconnection, Turbulence and Electrodynamics of the Moon's Interaction with the Sun
DEM	Digital Elevation Model—a digital model for 3D presentation of a terrain's surface, created from terrain elevation data
ESA	European Space Agency
GPS	Global Positioning System
GRAIL	Gravity Recovery and Interior Laboratory
ISS	International Space Station
JAXA	Japan Aerospace Exploration Agency
LAC	Lunar Aeronautical Chart
LADEE	Lunar Atmosphere and Dust Environment Explorer
LCROSS	Lunar Crater Observation and Sensing Satellite
LRO	Lunar Reconnaissance Orbiter
LROC	The Lunar Reconnaissance Orbiter Camera
NASA	National Aeronautics and Space Administration
QR code	Quick Response Code
SELENE	Selenological and Engineering Explorer (JAXA)
STEM	Science, Technology, Engineering, and Mathematics
VR	Virtual Reality
WAC	Wide Angle Camera

C. Zhang (✉)
NASA Scientific Visualization Studio at Goddard Space Flight Center, Greenbelt, MD, USA

The Ohio State University, Columbus, OH, USA
e-mail: cheng.zhang@nasa.gov

© Springer International Publishing Switzerland 2018
O. Ahlqvist, C. Schlieder (eds.), *Geogames and Geoplay*, Advances in
Geographic Information Science, https://doi.org/10.1007/978-3-319-22774-0_11

11.1   Introduction

The Moon is the only natural satellite of the Earth—the closest celestial object to our home planet. Appearing in folktales of almost all cultures, the Moon has fascinated people for thousands of years. Even though humanity has accomplished great achievement of lunar exploration, only 12 people have thus far landed on the Moon; It remains inaccessible for the rest of us. To rectify this, **The Moon Exploration** is a geocaching, multiplayer, mixed reality game that brings the Moon down to the Earth so that people can have access to it. Using the most up to date scientific data, players can explore in the game as if they were the astronauts exploring on the Moon. The location-based mapping scheme maps a lunar location to places on the Earth, so people can explore the Moon in the virtual world while moving around on the Earth. The game-playing facilitates communication and social interactions among players, which could facilitate the formation of a large lunar geocaching community on the Earth.

Since 2000, the scientific community has seen a new trend of private sector and even individuals participating in space exploration. This grass-roots movement needs an accessible platform where everyone can try out and feel what things would be like on the Moon. **The Moon Exploration** platform can provide universally accessible and useful information about the Moon. The game uses data from lunar landing missions, spacecrafts, satellites, or lunar orbiters. The goal is to bring the Moon down to the Earth in an appealing way so that people can explore the Moon as if they were the astronauts in the virtual world while moving around on the Earth. The pervasive play has the potential to stimulate young generations' interests in space exploration and promote STEM learning. The game can also be a platform for potential citizen science projects in the future, which would benefit the science research community as well.

In the following section, I offer an overview of lunar exploration and lunar geology. I also review the geocaching games and the technology of virtual reality (VR) and augmented reality. I describe how the game is designed and developed in Sect. 11.4.

11.2   Background

People have studied the Moon since ancient times. Ancient astronomers understood the lunar regular cycle of phases, the lunar gravitational influence on the ocean tides, the cause of lunar eclipses, and could even predict solar eclipses by analyzing the Moon's motion (Zhentao et al. 1989; Steele and Stephenson 1998; Stephenson 1997). It was Galileo Galilei who first observed lunar mountains and craters when he pointed his crude telescope to the Moon. Nevertheless, it is only in modern times that human beings could actually explore the Moon.

11.2.1 The Exploration of the Moon

Since 1959, 78 spacecraft have so far (to date) orbited, impacted, flown by, or landed on the Moon (Lunar and Planetary Institute 2016; Williams 2013; Spudis 2008). These missions can be divided into two phases: the lunar exploration driven by the space race from 1950s to 1976 and the current exploration that began in the 1990s. In the first phase, both competitors the Soviet Union and the United States have accomplished great achievements.

From the late 1950s to 1976, the Soviet Union carried out the Luna Program, a series of robotic spacecraft missions to the Moon. On January 2, 1959, Luna 1, the first robotic spacecraft to the Moon, missed its intended impact and became the first spacecraft to fall into orbit around the Sun. In September 1959, Luna 2 was launched and successfully hit the Moon's surface, becoming the first man-made object to reach the Moon. Launched in October 1959, Luna 3 returned the first photographs of its far side, which can never be seen from the Earth. Luna 9 was the first spacecraft to perform a successful lunar soft landing in 1966. Luna 10 was the first artificial satellite of the Moon. Luna 16 in 1970, Luna 20 in 1972, and Luna 24 in 1976 returned a total of 0.3 kg rock and soil samples from the Moon. The Soviet Lunokhod program landed two pioneering robotic rovers on the Moon in 1970 and 1973.

On the other hand, The United States explored the Moon in two steps: (1) the robotic spacecraft missions to pave the way for manned landing; (2) the Apollo program landing men on the Moon. The first step had three missions: six hard landing Ranger missions (1961–1965) (Williams 2005b; Hall 1977), five Lunar Orbiters missions (1966–1967) (Lunar and Planetary Institute 2011a), and the Surveyor space probes (1966–1968) (Lunar and Planetary Institute 2011b). Meanwhile, the Apollo Program (Loff 2015) was developed in parallel. Apollo 8 made the first crewed mission to lunar orbit in 1968. The Apollo 11–17 except for 13 (1969–1972), six successful manned landings are seen as the culmination of the space race. The Apollo missions return about 382 kg of lunar rocks and soil in about 2200 separate samples. Many scientific instrument packages were installed on the lunar surface. For example, the stations' lunar laser ranging corner-cube retroreflector arrays are still being used. Starting in the 1990s, many other countries became directly involved in lunar exploration. Japan (NASA 2007; JAXA 2007), China (Huixian et al. 2005; Li et al. 2015), and India (Bhandari 2005; Pieters et al. 2009) have developed their lunar exploration projects and launched spacecraft to the Moon. Meanwhile, the United States has orchestrated several lunar exploration missions—Clementine (Dino 2008; Williams 2011), Lunar Prospector (Williams 2005a; Lunar and Planetary Institute 2010), LRO and LCROSS (NASA 2015), ARTEMIS (Folta and Woodard 2010; NASA 2013a), GRAIL (NASA 2011a, b) and LADEE (NASA 2013b, c) for further information on the Moon such as gravitational fields, the chemical composition and distribution of the moon surface, and the Moon's interaction with the Sun.

11.2.2 Lunar Geology

The knowledge gained from lunar exploration helps us better understand early history of the solar system and the lunar geology (Wilhelms 1987; Jolliff et al. 2006). Unlike the Earth, the Moon lacks a significant atmosphere and does not have a dipolar magnetic field. Its radius is only about one fourth of the Earth, and its gravity is only about one sixth that of the Earth. The crust of the Moon is on average about 50 km thick.

The lunar surface is mainly covered by bright and dark areas. Lighter areas are lunar highlands and darker areas are maria. The highlands are older than the visible maria, and hence are more heavily cratered. The maria are the major products of volcanic processes on the Moon. At a close look, the lunar landscape is characterized by craters, volcanoes, mountains, valleys, rilles, domes, wrinkle ridges, and grabens. According to the study by Head et al. (2010), there are 5185 craters larger than 20 km in diameter. Impact cratering is the most noticeable geological process on the Moon. The largest impact basins were formed during the early periods. They were successively overlaid by smaller craters. Small craters tend to form a bowl shape, whereas larger impacts can have a central peak with flat floors. These lunar geological data would be embedded into the virtual game world.

Elements presented on the lunar surface include oxygen, silicon, iron, magnesium, calcium, aluminum, manganese and titanium. Rare elements such as titanium, gadolinium, and terbium draw growing interest of the Moon (SPACE.com 2011; David 2015). Another potential treasure on the moon is Helium-3, an isotope that may support fusion in the future. Helium-3 is a component of the solar wind. Since the moon doesn't have much atmosphere, scientists estimate that there may be more than a million tons of isotope on the Moon.

11.2.3 The New Trend and Associated Challenges

As of 2000, space exploration is no longer exclusively the realm of large government-driven programs. The private sector has increasingly play a role—sometimes even a critical one. For example, as of 2015 SpaceX has flown six missions to supply the ISS with cargo. NASA also awarded SpaceX a contract to develop a program to transport crew to the ISS. The important feature of this new trend is the affordability of access to space. Because of the low cost and the readiness of technologies, more grass-roots start-up companies, or even individual projects have appeared. For example, teams in Google Lunar XPrize[1] (XPrize Foundation 2007) are all privately funded, and some have started from an individual's idea. This trend also forms the

---

[1] Google Lunar XPRIZE is a $30 million competition to land a privately funded robot on the Moon. Google Lunar XPRIZE aims to open a new era of lunar travel by vastly decreasing the cost of access to the Moon and space.

collaborations between large government space agencies and the private sector. Many private companies or individuals seek launch vehicles (usually operated by large government programs) to lift off into space as payloads or space tourists (Sven 1996; Space.com 2007; Fox 2010). NASA has developed many programs to benefit from the commercial attempts to space. For example, in 2010, NASA awarded six companies the ILDD (Holmer 2012) contracts for the purchase of technical data resulting from industry efforts to develop vehicle capabilities and demonstrate end-to-end robotic lunar landing missions (Braukus et al. 2010). The data from these contracts will inform the development of future human and robotic lander vehicles and exploration systems.

Recruiting the general public (individual amateurs) as citizen scientists to conduct serious research is another example of the trend. Zooniverse (Zooniverse 2013) is a citizen science web portal produced and operated by the Citizen Science Alliance (Citizen Science Alliance 2013). Hosted by Zooniverse, the Moon Zoo (Galaxy Zoo 2010) is a citizen science project that asks users to identify, classify, and measure the shape of features on the lunar surface by providing released Planetary Data System (PDS) with high spatial resolution images from NASA's Lunar Reconnaissance Orbiter (LRO) camera instrument. So far the Moon Zoo users have already visually classified 3,915,560 images and help researchers study the lunar surface in unprecedented detail (Katherine et al. 2010) and address important themes of lunar science and exploration.

In future physical lunar exploration, the goal is beyond landing spacecraft or people on the Moon surface. It is more about how to make use of the lunar resources and how to build a hub or colony (Schmitt 2005). On November 26, 2015, President Obama signed the U.S. commercial Space Launch Competitiveness Act (or H.R. 2262) into law (114th Congress 2015). The law grants companies the rights to the natural resources that they mine from outer space, including Moon, Mars, asteroids, and other heavenly bodies. In 2013, ESA teamed up with architect companies (such as Foster and Partners) to test out various Moon base-building technologies including 3D printing. They concluded that 3D printing using lunar soil was feasible in principle for constructing buildings and other structures (3DERS 2014). Recently, ESA announced its plan to build a 3D printed Moon Village by 2030 in the International Symposium on Moon 2020–2030 in the Netherlands. The new Moon Village is designed to replace the International Space Station (ISS), and could provide a potential springboard for future missions to Mars. According to the estimation by NexGen Space LLC, a consultant company for NASA, a lunar refueling station would reduce the cost of sending humans to Mars by as much as $10 billion per year (Crew 2016). A moon village seems to look pretty inevitable.

Given the above achievement, people should have a better chance to reach the Moon. However the reality suggests otherwise. So far only 12 Apollo astronauts have landed on the Moon since the Apollo program began. The Moon still remains unreachable for most people on the Earth. The Moon can be regarded as the eighth continent of the Earth, but given its distance of 384,400 km from the Earth, it is a long journey for people to get to the Moon. It took Apollo Astronauts 3 days 3 h 49 min to reach the Moon in 1969. By far the fastest mission to fly past the Moon was NASA's New Horizons Pluto mission. The powerful rockets propelled it over 58,000 km/h speed. Still, it took 8 h and 35 min to reach the Moon (Williams 2016). Even for the elite, rich space tourists[2] (Fox 2010) without financial concerns, space travel is still a daunting task physically and mentally, which also requires people to obtain systematic knowledge and rigorous trainings. The Moon is a remote and uncharted territory. It's a lifeless, rainless, sun-seared, barren, and hostile world. Survival in such an extreme environment is difficult for ordinary people. Given the cost, training, physical conditions, space tourism is still not accessible for the general public. While available, large scientific data of the Moon are most likely used by and interpreted for science community rather than the general public. Because the original data are usually in the raw format of instruments and data in different processing levels (NASA 2010) may be involved with various parameters (or in different map projections (Fenna 2006; Snyder 1987)), they would be too complicated for the general public to comprehend. The challenge is how to access the Moon with the first-hand experience in an easy, interesting, and safe way with low cost.

Through computing technology, people can have access to more information about the Moon than before. For example, NASA has produced a lot of lunar data products including latest lunar topography with the resolution of half meter per pixel. NASA Scientific Visualization Studio (Studio NSV 2000) has created a lot of animations about the Moon such as lunar phases, libration, eclipses, evolution, craters, and other features which precisely convey the latest scientific achievements to the general public. However, all these animations are piece by piece, linear in storyline. They cannot provide users with interactive and immersive experience of being on the Moon. Google Moon (Google 2012) allows viewers to have a virtual tour of the Moon and visit Apollo landing sites, but the number of sites that can be visited is very limited. Moon Zoo can provide users with a lot of 2D high-resolution lunar images and show the live information of different users online. However, users cannot interact with each other in real time. In other words, the current available applications/software lack an integrated solution that provides users with interactive, immersive, and first-hand virtual experience on the Moon, which is very much needed as a sandbox for the future physical lunar exploration—lunar mining and constructions of permanent lunar bases.

11.3   Related Work

Besides the rich lunar content, **The Moon Exploration** also touches two exciting areas, geocaching games and VR/AR technology. In this section, I offer two short overviews.

---

[2] Space tourism has emerged since 2000. Seven rich individuals spending $20 million to $40 million have become space tourists.

11.3.1 Geocaching Games

To bring down the Moon as the eighth continent of the Earth, I found that **geocaching** is an appealing idea to make it work. Originally from a very old treasure-hunt game, geocaching has been around since 2000 when President Clinton announced the removal of Selective Availability of GPS (FAA 2000). Geocaching is a GPS-enabled, location-based treasure hunt. In this outdoor recreational activity, participants (geocachers) can hide or hunt 'treasure' (geocache or cache) by using GPS receivers. A geocache usually consists of a waterproof container that contains a logbook and inexpensive items. The coordinates of geocaches are posted onto a geocaching website. When a geocacher finds a geocache, he records the date and signs the logbook with his established code name. Afterwards, the cache must be placed back in the exact location where it was found.

Since 2000, geocaching has rapidly become popular worldwide. Many geocaching companies have been established, for example geocahing.com, NaviCache (NaviCache 2011), TerraCaching (TerraCaching 2004), and **Munzee** (Munzee 2012). Geocaching is now played in more than 200 countries around the world, and there is at least one physical geocache deployed on every continent, including Antarctica (Spencer 2012). **Ingress** is another location-based game created by Niantic Inc., previous under Google[3]. **Ingress** is not exactly geocaching but has very similar game structures as a location-based game. The game was first released in November 2012. In 2015, **Ingress** already has seven million players worldwide. The hottest location based game to date, Pokémon Go, released by Niantic in July 2016, quickly became a global phenomenon. It has reportedly been downloaded by more than 100 million people worldwide even with mixed review

In their work (Farvardin and Forehand 2013), Farvardin and Forehand conducted a survey about the motivation of geocaching, which reveals the reasons why the game is so appealing. The top motivations include (1) The thrill of the hunt, the opportunity to discover and explore new places, (2) Natural wonders and memorable experiences, (3) A way to get exercise, (4) Challenge, Solving puzzles, Cracking difficult codes (5) Socializing, Meeting new people, Making friends; Likeminded folks. These are also the motivations that we will try to integrate into our solution.

11.3.2 Virtual Reality and Augmented Reality

Since Ivan Sutherland and Bob Sproull created the first head mounted display (HMD), Sword of Damocles in 1968 (Sutherland 1986), VR/AR has become an exciting area in academia, military training, and medical fields. Some fundamental work has been done, particularly in theory. The concept the reality-virtuality continuum proposed by Milgram et al. (1994) indicates no clear line between reality

---

[3] Ninantic Inc. spun off from Google as an independent company in August 2015.

Fig. 11.1 Some affordable HMDs in the current market. Goggle Cardboard, Samsung Gear, and Goggle Tech C1-Glass can be directly used with cell phones. Goggle Tech C1-Glass can be folded into a glass case. With 70 sensors and two wireless infrared cameras (not shown here), HTC Vive Pre offers users better motion tracking while users are moving freely around in a room. Fove is the first eye-tracking HMD. Microsoft HoloLens is expensive but has built-in Windows 10. It can merge real-world elements with virtual holographic images, creating half virtual and half augmented reality

and virtual reality. In his forward-looking book (Zhai 1998), Zhai presented a philosophical view of virtual reality as invented realities that may be more real than people dare to believe.

For nearly 50 years, even though many interesting developments have been achieved, VR/AR industries are still in smoldering stage. Several advances in technology have made the widespread use of VR more possible than before. One is the development of low-cost high-quality mobile devices and another is wearable computing, which makes affordable and comfortable VR devices available in the market as shown in Fig. 11.1. Google's Cardboard, which costs about $20, helped introduce VR to mainstream consumers. Consumers seem to be ready for VR. Samsung's $99 Gear VR sold out on Amazon during the 2015 holiday season within 2 days. Most mobile phones can work with Google's Cardboard and Samsung's Gear VR, which allows people to have immersive experiences at any time anywhere. Meanwhile, enterprises are beginning to see the potential of VR applications such as gaming, entertainment, training, manufacturing, communication, and so on. According to a report from Digi-Capital, both VR and AR markets will become mainstream by 2020, and will generate $150 billion in the next 5 years (Gaudiosi 2015). The large IT companies such as Google, Facebook, Samsung and Microsoft all have products in the virtual reality space. In a research report from Goldman Sachs, the virtual and augmented reality market could become an $80 billion industry by 2025. Even though there is a big gap between these two reports, both agree that VR reaches far beyond gaming and entertainment. Gaming and entertainment will still drive much of the growth, but car makers, retailers and even interior designers could bank on VR technology, according to Goldman Sachs analyst Heather Bellini.

The affordable and portable VR devices such as those in Fig. 11.1 are the essential component to **The Moon Exploration**.

11.4   The Approach

How do we make the Moon easily accessible? Can we access the Moon as easily as we access a place on the Earth via Google Maps or Google Earth? We design our solution as a multiple-players, geo-caching, mixed reality game. The goal is to create an effective learning experience in an interactive virtual lunar world with real world Global Positioning System (GPS) data. The main idea is to use real world GPS data to make a connection between a location on the Moon and places on the Earth. The link through GPS data shortens the distance between the Earth and the Moon, which makes the Moon become the accessible eighth continent of the Earth. Treasure-hunting is a proven appealing game genre that has existed for thousands of years. **The Moon Exploration** integrates treasure hunting as inviting activities for players. As a result, players can explore on the Moon surface in the game world while they are actually on the Earth.

Behind the scenes, **The Moon Exploration** has two main game mechanics, the two module structure and the mapping scheme. The two-module structure is designed for the game to grow efficiently and flexibly. The mapping scheme makes use of both geo-data of the Moon and the Earth to bring the Moon down to the Earth. The two-module structure and the mapping scheme are presented in Sects. 11.4.1 and 11.4.2. In terms of locale, **The Moon Exploration** can be divided into two parts, the Moon (the virtual world) and the Earth (the real world). The lunar virtual world is created in the computer with Unity3D based on the available scientific data mainly from Lunar Reconnaissance Orbiter (LRO). LRO's standardized lunar coordinate system (NASA GSFC 2008) is used to determine any location on the Moon surface with a unique set of latitude, longitude, and elevation. The coordinate system is also employed to compute the accurate positions and movements of the Sun and the Earth at runtime. The virtual world also contains manmade objects such as Lunar Module, Lunar Roving Vehicle (LRV or Moon Buggy), science equipment, and virtual caches. Each player has his own avatar (as in Fig. 11.2) in the virtual world as well. More details of the virtual world are in Sect. 11.4.4. In the real world, the game setting and playing are not much different from other online games. Players can create a cache, place it in a location, explore the lunar world, or hunt and find a cache. Several modes (creation mode, adventure mode, spectator mode, and multi-player mode) are used to facilitate various play activities. The game also provides players with the option to plug in virtual reality devices such as **Oculus Rift**, allowing players to have first-hand 3D immersive virtual experiences. The gameplay is described in Sect. 11.4.3.

Fig. 11.2 The default
avatar is created based on
the appearances of Apollo
astronauts

11.4.1 Two-Module Structure

Involving the two worlds, the Moon and Earth, the content of **The Moon Exploration** can be very broad and rich. The usage of the game may evolve further in various directions over time. For it to become more efficient and flexible, I designed an open-end two-module structure shown in Fig. 11.3. One module is for end players, the other for developer (expert) team. An approval process is needed to connect the two modules. This process is to ensure that the entire two-module structure works.

In the player module, players have two options for their game property, public and private. The game property of a player refers to the player's playing data, ideas, designs, and new implementations of the game. Under the public option, a player can make his/her game property available for the approval process toward being published and eventually available for all players once approved. Under the private option, a player can keep his/her game property private from the public.

In the expert-team module, the team has the authority of further developing the game. For example, the team can create and approve new developments, modify simulation models, configure the general game settings, and analyze and validate the collected data. In the approval process, players provide the inputs. The experts can have access to and evaluate the inputs. Qualified ones will be integrated into the system and become available to all users. Unqualified ones will be filtered out and only be available for the submitter himself. This structure allows the game to evolve as new developments (from both players and developer team) become available. Having users' inputs as a part of new development is a very powerful approach to enhance the existing application. For example, Google Maps allows users to add new features or edit existing ones with pending approval (Google n.d.). A user can

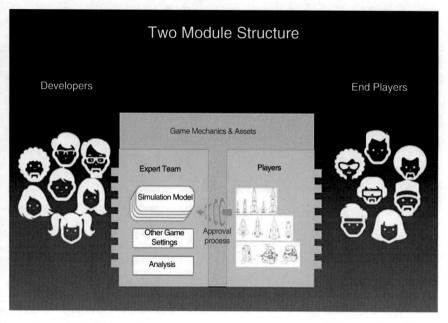

Fig. 11.3 The two-module structure

add new points of interest such as restaurants, hospitals, schools. A user can also add a walking trail, biking path, and more. Based on individual knowledge and experience, a lot of fine details can be added into Google Map by users via the approval process. In return, Google Maps can provide users with meticulous information on places.

This architecture can also turn the game into a powerful crowdsourcing platform where players' data, design, and implementation can be available and analyzed. In this setting, research such as citizen science projects can be carried out. **The Moon Exploration** is an open world game that has no single final goal for players to accomplish, providing players with a large amount of freedom in choosing how to play the game. However, the game has an achievement/scoring system to motivate players to gain recognition for their performance in their published profiles.

Other than the two-module structure, mapping a lunar location to places on the Earth plays a key role in dragging the Moon to accessible range. In the next section, the mapping scheme is described in detail.

11.4.2   The Mapping Scheme

The mapping scheme is a critical component that fastens the Moon and the Earth together smoothly, which allows people to explore on the Moon while moving around on the Earth. The mapping scheme is based on the two coordinate systems, one for the Moon and the other for the Earth. Any location on the moon or on the

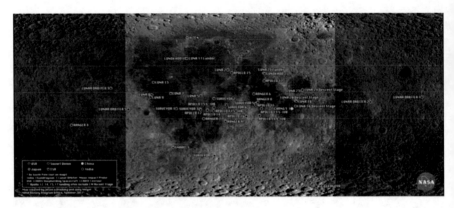

Fig. 11.4 The map of all artifacts on the Moon. Source: NASA

Earth can be specified by a set of numbers such as latitude, longitude, and elevation in these coordinate systems. For the Moon, I use LRO's standardized lunar coordinate system (NASA GSFC 2008), which is comparable to the latitude and longitude of Earth in the common geographic coordinate system for the Earth.

Given a lunar location with its latitude and longitude, we can find the corresponding position on the Earth with the same values of latitude and longitude. However, in this naïve mapping, most points of interest, such as robotic or manned lunar landing sites and the locations of most manmade artifacts left on the near side of the Moon (see Fig. 11.4) would be mapped into somewhere in oceans, desserts or jungles in Africa. In a reverse mapping, The United States is mapped into places on the far side of the Moon. Another issue is about the scale. The lunar diameter is about 3474 km, 27.242% of the Earth diameter. The total area of the lunar surface is only 7.4% the Earth surface area. In fact, the scale directly affects the speed of exploration in the virtual world. If 1:1 scale ratio is used, the United States would take up almost half of the Moon (see Fig. 11.5). The scale ratio 0.27242:1 would make more sense in the following situation. If the driving speed is 60 mph on the Earth, then the corresponding driving speed would be about 15 mph on the lunar surface, which is closer to the record of lunar rover vehicle (LRV) by Eugene Cernan, 11.2 mph (18.0 km/h). LRVs were designed with a top speed of about 8 mph (13 km/h) in the Apollo program.

In our solution, scale ratio is a variable that can be changed by a player to fit different scenarios. When players interact with each other, their scale ratios need to be the same. The mapping function can be determined by several parameters including the initial lunar location $P_{t_0}^L$ and the start point on the Earth $P_{t_0}^E$.

$$P_{t_0}^L \leftrightarrow P_{t_0}^E \tag{11.1}$$

$$P_{t_i}^L = P_{t_{i-1}}^L + \vec{d} \cdot \Delta t \cdot s \tag{11.2}$$

Fig. 11.5 The U. S. mapped on the near side of the Moon at scale 1:1, courtesy of Nicholas Schroen

When a player chooses a lunar location he wants to visit, his initial position on the Earth is mapped to the selected lunar location as defined in Eq. 11.1 with a mapping scale s. Afterwards, anytime his lunar location can be computed through Eq. 11.2, where $P^L_{t_{i-1}}$ is the previous lunar location, $\vec{d}$ is the direction vector, Δt is the time step, and s is the scale ratio. Note that the direction vector $\vec{d}$ is measured by the player's movement on the Earth if he's playing in the real world with GPS turned on in his device. However, the direction vector $\vec{d}$ can be computed by the movement of the player's avatar on the lunar virtual world if the player is indoors or his device has no GPS signals.

The mapping is one-to-many from the virtual world to the real world. For example, the point (Lon: 30.77°, Lat: 20.19°) in Taurus-Littrow Valley (Apollo 17 landing site) on the Moon can be mapped to the location (Lon: −76.851531°, Lat: 38.996078°) in NASA Goddard Space Flight Center, the location in Time Square (Lon: −73.98513°, Lat: 40.758896°) in New York city, or the location (Lon: 2.338568°, Lat: 48.860371°) at the Louvre in Paris. This one-to-many mapping allows people in different places on the Earth to explore one point of interest on the Moon.

Similarly, caches follow the same scheme. A virtual cache can have many corresponding real caches on the Earth. The score calculation reflects this mapping scheme—a player will achieve the maximal score for seeking a cache if he can find the virtual cache and all its corresponding real caches. Previous lunar explorations left many artifacts (NASA 2011c) on the lunar surface, which can be ideal candidates for super caches (see Fig. 11.4). The details about cache design is in Sect. 11.4.4.

11.4.3 Gameplay

The Moon Exploration is a mixed reality game. Its play is pervasive, extended out in the real world. Via gameplay, people can explore the Moon in the virtual world while moving around on the Earth. On the other hand, a player can also play the game without stepping out of the door, that is, no GPS data is needed. To have a proper game environment for players, I designed flexible options of play mode, mapping scheme, transportation tools, avatar, views, and so on. What's most important, there are four play modes available in the game.

1. **Creation mode**—assists a player in creating and building items such as caches, or anything as a part of a lunar colony. The Moon is believed to be a central hub for humans' further journey into space. Colonizing on the Moon is one of the important steps to explore the cosmos (Schmitt 2005). ESA, working with the renowned architectural company Foster and Partner, proved that 3D printing using lunar soil was feasible in principle (3DERS 2014). The game borrows this idea and allows players to have 3D printers to manufacture parts, then assemble them into a habitable base on the Moon, especially in lunar lava tubes (CNN 2010) or craters in lunar poles, such as Shackleton Crater at the South Pole.
2. **Adventure mode**—is designed for players to explore the Moon and find caches. As mentioned above, caches can be either virtual or real. When a player finds a cache, he should record the event in the log. If the cache is virtual, the system will automatically update the game status. If the cache is real (the player plays game in the real world), the player will use the cache ID to update the game status in a device. The player has an option to take an item from the cache and place a new item in the cache as well. For transportation, a player can have three ways to traverse on the Moon: walking, driving the buggy, and flying a small aircraft (as the transportation tools). More advanced tools require certain training and skills with higher cost. The game provides a menu to allow a player to choose his way to travel. If the speed does not fit in the player's choice, the game will display the menu window for the player to make a new choice.
3. **Spectator mode**—allows players to explore on the moon surface and watch other players playing activities such as hiding or hunting caches, or interacting with others. In this mode, a user can check and study certain features at a location on the moon just like the way we use Google Maps.
4. **Multiplayer mode**—enables multiple players to interact and communicate with each other on the virtual lunar world. Except for communication, players need to

have the same scale ratio to interact with each other. There are several ways for players to communicate with each other. Each player has his profile that is publicly available, but the player can turn off certain items in his profile. The game has a broadcasting board that everyone can see. A player can send a message for assistance or recruit his partner or team over there. A player can also send a private message to certain players if he knows their usernames. Since players would explore in the extreme, remote, harsh lunar territory, a positive and healthy relationship among players is critical to have an effective experience in pursuing the common goal. Jenova Chen's **Journey** has set up an excellent example of how to forge healthy companionships among players (That Game Company 2012a, b). The basic idea for **Journey** designed by Chen and his team, was to create a game that moved beyond the "typical defeat/kill/win mentality" of most video games. **Journey** is intended to make the player feel "small" and to give them a sense of awe about their surroundings (Nava 2012). In a similar situation where the Moon surface is a remote, harsh, and lifeless environment, players need solid friendships to achieve common goals in **The Moon Exploration.**

The four play modes are not mutually exclusive. Some can work together. For example, with both the creation mode and multiplayer mode on, a player can build a project with a partner or team. In addition, GPS tracking function can be turn on/off anytime. In some sense, this option offers a switch among different worlds. If GPS is off, a player can play the game in a device like most computer games (indoor). With the GPS on, a player can play the game while moving around on the Earth. This option can even help players handle the obstacles that they encounter on the Earth. For example, if a participant is playing the game while walking until a big lake blocks his way, he can turn off GPS and continue to explore the Moon in the virtual world with his device; alternatively he can continue to traverse on the Moon while rowing a boat in the lake with GPS being turned on.

At the beginning, a player can choose the avatar he wants to play. The default avatar looks like an Apollo astronaut, shown in Fig. 11.2. The player can also change the scale ratio s in the mapping function. By default, the game is played in the first person view to have a better sense of immersion/presence, particularly with VR display devices[4]. But in certain situations, the third person view may provide better perceptions. A player can change it as well.

11.4.4 The Game World

The game world consists of the lunar world and the earth world. On the display screen, a user can swap these two worlds anytime during the game. Like Google Map, the view of the earth world is simple, indicates the positions of real caches around the location on the Earth. The lunar world, however, is complex, is generated

---

[4]Like most current computer games, **The Moon Exploration** can be played without a VR device.

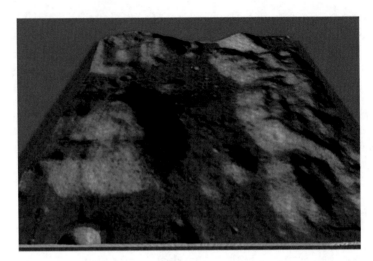

Fig. 11.6 3D model of Taurus-Littrow Valley is created in Unity 3D based on real topographic data

procedurally based on the initial location and the real lunar topographical data from previous or ongoing missions such as LRO. For example, shown in Fig. 11.6, a 3D model of Apollo 17 landing site, Taurus-Littrow Valley is created in Unity based on topographic data such as DEM. Small items or textures in the lunar world are also generated procedurally based on available data. For instance, boulders can be created according to the boulder density hazard maps generated from the Moon Zoo. The purpose is to create a believable 3D environment, that is, a simulation of the lunar world containing useful lunar geological information. With the mapped GPS location and time, the playing environment is also simulated with the accurate motions of Sun, Earth, and other celestial objects in the solar system, creating the correct impression of what it looks like on the Moon in a specific location at a certain time. A compass and a location indicator are built in with the option to show the 2D map of the region, the facing direction, the location's latitude, longitude, and the names of surrounding geological features such as mountains, craters, and maria, etc. Some other useful information such as the distribution/concentration of certain elements can also be embedded in the game.

The lunar world also contains virtually manmade materials—tools and equipment, constructions built by players with the approval of the expert team, the artifacts left by previous missions, and caches. Tools and equipment include transportation means such as rovers and aircraft. The higher the level of the players, the more skills they possess, and the higher-end transportation means are available to them. Similarly, the distribution of equipment such as 3D printers and various robots follows the same pattern. Note that 3D printers are the revolutionary protocol for manufactory. They can be used to create gears and parts of any large tools and equipment. The initially available robots help assemble them together. The process can also create more available robots and eventually lead to the creation of a colony.

In **The Moon Exploration**, caches are essential ingredients. Like most geocaching games, a cache is a container at a given coordinates either on the lunar surface or on the Earth. A cache must contain a logbook with a unique ID, owner's name and contact information. It may also include items for trade and other activities. When a player finds a cache, he must record his activity in the logbook (such as when and where the cache is found, and by whom, and whether any item is taken away or left out, etc.). To accommodate various activities, I have designed several types of caches: lunar cache, earth cache, and event cache.

1. A **lunar cache** is on the lunar surface. Its size may vary. Along with the logbook, it may also contain information of surrounding geological features, personal experience, hints, tips, instructions, or warnings. A player can leave/inform helpful messages for later visitors.
2. An **earth cache** is like a traditional geocache, but is placed at a coordinate on the Earth rather than in the virtual lunar world.
3. An **event cache** is about a gathering of players in the game community. The Event Cache specifies a time for the event and may provide coordinates to its location in the virtual lunar world or on the Earth. After the event has ended, it is archived.

Among caches, many interesting things can be arranged. For example, I designed a so-called chained cache—in order to completely recover a cache, a player may need to find out other caches to obtain critical information required by this cache. The player can broadcast to find a partner(s) to assist him. With this twist, communication and collaboration can be forged naturally.

11.4.5 *Players and Communities*

The Moon Exploration is a cultivated learning environment where players can have effective first-hand learning experiences. It is a multi-level game with an open-ended structure that can evolve as an effective crowdsourcing platform where serious lunar research can be conducted. It is also a sandbox where innovated designs or blueprints can be conceived, tested, and modified. Therefore, the target users are in a large range—from third or fourth graders all the way to graduate students, professionals, and experts. Players have several interaction options when playing the game. Every activity is possible for solo players but larger and more complicated tasks, for example, building a lunar base/colony, are more appropriate for groups. Players can form a virtual group, community, or society in the play world. The interactions among players help promote building healthy friendships and companionships in the harsh environment while pursuing common goals. As this game will be across-platform, it can be played in desktops, laptops, tablets, or mobile phones. **The Moon Exploration** would be the lunar information center, and can be made universally accessible and useful.

The game is designed as an open platform, in which the game can grow via not only the development team but also the player community. The game content can also be expanded to include more interesting, scientific, and educational materials. The complexity of **The Moon Exploration** requires (or draws) various skills and knowledge from different disciplines, which would lead to a multidisciplinary community. As the community grows, a virtual economy may also merge.

11.4.6 Development

The Moon Exploration is designed to be a multiplayer online geocaching game, which involves complex content of lunar science and complicated computing technology to support such a game system. To make the implementation feasible, I divide the whole project into at least three phases.

Phase one includes the implementation of characters/avatars, basic game world, interface and interaction design, the core game mechanics, and key algorithms. The key algorithms procedurally generate certain aspects of the game world. For example, I created an algorithm to generate lunar terrains procedurally in runtime based on the initial location and the lunar topographic database. The algorithm works with the terrain engine of Unity3D to create lunar terrain so that it would never run out in play. Similar algorithms will be implemented, for instance, to generate boulders on the lunar surface based on geological feature data.

Major tasks in phase two include (1) expanding the game to handle big geo-data, e.g., how to smoothly transit among different data sets; (2) building the game into different platforms—PCs, smart phones and handheld tablet computers; (3) making the game available in a local network with multiplayers.

Scaling up the game with hundreds of players online is the challenge in phase three. A well-known issue is to develop the engines to handle vast numbers of players. The engines are made of powerful servers. If a typical server can handle a certain number of players simultaneously, dividing the players into many servers proves to be an efficient solution. Another predicted difficulty is time synchronization across hundreds or thousands of players since time is used to drive many physics simulations as well as scoring and damage detection.

In phase one, I use the index map with 144 quadrangles (shown in Fig. 11.7) as the basic structure in my algorithms. The map with 144 plates, adapted from Bussey and Spudis' work (Bussey and Spudis 2012), corresponds to the widely used Lunar Astronautical Chart (LAC) series. Each plate has 2° overlay with its adjacent plates if any, which ensures a smooth transition when traversing from one quadrangle to another. In terms of location, lunar terrains, data, game items can be efficiently organized with this structure. Given a location that a player wants to visit, it is easy to determine which quadrangle the location is in. Then the game can load all the information and objects related to that quadrangle at runtime while ignoring the rest. For each quadrangle, data are hierarchically organized from low resolution to

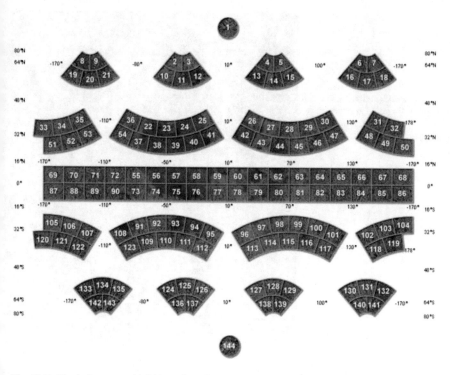

Fig. 11.7   The index map with 144 quadrangles corresponds to the LAC system. At the lunar equator a plate is 20° in longitude and 16° in latitude. Source: NASA

high resolution like a pyramid. Based on the distance between the camera and the destination in the 3D scene, the algorithm can automatically choose the proper resolution data to be used.

When creating lunar terrain, a common issue is the missing data in a high resolution file as shown in Fig. 11.8, I create an algorithm to repair the holes. The general idea is to make use of the hierarchical data in various resolutions. If data are missing in a high resolution file, then the algorithm interpolates the corresponding value from surrounding pixels in the same file or interpolate it from lower resolution data. Since topographic data (usually in DTM) (Tran et al. 2010) are often associated with a confidence map that gives users a guidance on what elevation values to trust and not to trust, we can easily find out which pixels have missing data. Then the algorithm calculates the missing values through various resolution data files.

To make lunar information universally accessible and useful, **The Moon Exploration** is designed as an across-platform game. It can be played in desktops, laptops, tablets, and mobile phones. The game also provides users an option to connect their VR display devices such as Oculus Rift, Google cardboard, etc. during the game-play, which would provide players a believable and immersive experience.

228   C. Zhang

Fig. 11.8 Topographical
data of Apollo 17 landing
site (Taurus-Littrow
Valley) is available at
LROC website http://lroc.
sese.asu.edu/. The
resolution is 2 m per pixel.
The missing data appears
in the middle of the image

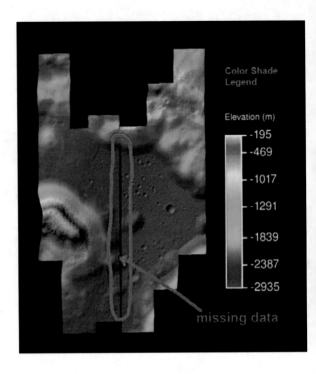

11.5   Summary

The Moon Exploration is a multiplayer online geocaching game based on real scientific data from direct lunar exploration, space crafts, satellites, and observations on the Earth. The goal is to make the Moon accessible as the "eighth continent" of the Earth. A user can explore any place on the lunar surface in the virtual world while moving around on the Earth. The gameplay provides a participant believable first-hand experiences as if they were the astronauts exploring on the Moon. The gameplay also fosters communication and social interactions among players to form a community of lunar geocaching on the Earth. **The Moon Exploration** is designed as a powerful crowdsourcing platform in which serious lunar research can be conducted. It may serve as a sandbox where innovative ideas, designs, and plans can be prototyped, tested, and modified. Finally, the game is also meant to attract the young generations' attentions to space exploration and STEM learning.

Acknowledgements I am very grateful to the Editors and two anonymous reviewers for their constructive and insightful comments. I would also like to thank my colleague Ernest Wright for his proof reading and comments that greatly improve the exposition of the manuscript. All remaining errors are my own.

References

114th Congress (2015) H.R.2262—U.S. Commercial Space Launch Competitiveness Act. https://www.congress.gov/bill/114th-congress/house-bill/2262/text# toc-HEE062BAAFBDD 4A43859C0142C68E67F9. [Online]. Accessed 30 Dec 2015

3DERS (2014) ESA outlines plan to use 3D printing to build a fully habitable base on the Moon. http://www.3ders.org/articles/20141107-esa-outlines-plan-to-use-3d-printing-to-build-a-fully-habitable-base-on-the-moon.html. [Online]. Accessed 20 April 2015

Bhandari N (2005) Chandrayaan-1: science goals. J Earth Syst Sci 114:701–709. https://doi.org/10.1007/BF02715953

Braukus M, Madison L, Byerly J (2010) NASA awards contracts for innovative lunar demontrations data. https://www.nasa.gov/home/hqnews/2010/oct/HQ_10-259_ILDD_Award.html

Bussey B, Spudis PD (2012) The Clementine atlas of the Moon. Cambridge University Press, Cambridge, UK

Citizen Science Alliance (2013) A community of practice for the field of public participation in scientific research. http://citizenscienceassociation.org/

CNN (2010) Moon hole might be suitable for colony. http://www.cnn.com/2010/TECH/space/01/01/moon.lava.hole/

Crew B (2016) European Space Agency announces plans to build a 'Moon village' by 2030. [Online]. Accessed 15 Jan 2016

David L (2015) Is moon mining economically feasible?. http://www.space.com/28189-moon-mining-economic-feasibility.html. [Online]. Accessed 11 Nov 2015

Dede C (2007) Reinventing the role of information and communications technologies in education. In: Smolin L, Lawless K, Burbules N (eds) Information and communication technologies: considerations of current practice for teachers and teacher educators [NSSE Yearbook 2007 (106:2)]. Blackwell, Malden, MA, pp 11–38

Dino J (2008) NASA—Clementine mission. http://www.nasa.gov/missionpages/LCROSS/searchforwater/clementine.html. [Online]. Accessed 21 Oct 2015

FAA (2000) Satellite navigation—GPS—Presidential policy. https://www.faa.gov/about/office_org/headquarters_offices/ato/service_units/techops/navservices/gnss/gps/policy/presidential/. [Online]. Passed 21 Aug 2015

Farvardin A, Forehand E (2013) Geocaching motivations. Worcester Polytechnic Institute, Worcester

Fenna, D. (2006) Cartographic science: a compendium of map projections, with derivations. CRC, Boca Raton

Folta D, Woodard M (2010) ARTEMIS—the first Earth-Moon libration orbiter. https://www.nasa.gov/mission_pages/themis/news/artemis-orbit.html. [Online]. Accessed 11 Oct 2015

Fox S (2010) Private companies that could launch humans into space. https://www.space.com/8541-6-private-companies-launch-humans-space.html

Galaxy Zoo (2010) Moon zoo. https://www.zooniverse.org/

Gaudiosi J (2015) VR, AR will generate $150 billion in the next five years—Fortune. http://fortune.com/2015/04/25/augmented-reality-virtual-reality/. [Online] Accessed 10 Jan 2016

Google (2012) Google Moon. https://www.google.com/moon/. [Online]. Accessed 21 June 2015

Google (n.d.) Google map maker. https://www.google.com/mapmaker. [Online]. Accessed 15 Jan 2016

Hall RC (1977) Ranger impact—a history of project ranger. Tech. rep. NASA SP 4210. National Aeronautics and Space Administration. http://history.nasa.gov/ SP-4210/pages/Cover.htm

Head JW III, Fassett CI, Kadish SJ, Smith DE, Zuber MT, Neumann GA, Mazarico E (2010) Global distribution of large lunar craters: implications for resurfacing and impactor populations. Science 329:1504–1507

Holmer CI (2012) An overview of the Innovative Lunar Demonstration Data (ILDD) program: NASA's next steps to extending public/private partnerships beyond earth orbit. In: Lunar and planetary science conference, Lunar and Planetary Inst. Technical report, vol 43, p 1605

Huixian S, Shuwu D, Jianfeng Y, Ji W, Jingshan J (2005) Scientific objectives and payloads of Chang'E-1 lunar satellite. J Earth Syst Sci 114:789–794. https://doi.org/10.1007/BF02715964

JAXA (2007) KAGUYA (SELENE)—TOP. http://www.kaguya.jaxa.jp/index e.htm. [Online]. Accessed 21 May 2015

Jolliff BL, Wieczorek MA, Shearer CK, Neal CR (eds) (2006) New views of the Moon, vol vol. 60. Mineralogical Society of America, Washington, DC. http://www.minsocam.org/msa/rim/rim60.html

Katherine J, Crawford I, Grindrod P et al (2010) Moon Zoo: citizen science in lunar exploration. Astron Geophys 52(2):10–12

Li C, Liu J, Ren X, Zuo W, Tan X, Wen W, Li H, Mu L, Su Y, Zhang H, Ouyang JYZ (2015) The Chang'e 3 mission overview. Space Sci Rev 190:85–101

Sarah Loff (2015) The Apollo missions—NASA. http://www.nasa.gov/mission pages/ apollo/missions/index.html. [Online] Accessed 11 Oct 2015

Lunar and Planetary Institute (2010) Lunar prospector mission. http://www.lpi.usra.edu/lunar/missions/prospector/. [Online]. Accessed 21 Oct 2015

Lunar and Planetary Institute (2011a) The lunar orbiter program. http://www.lpi.usra.edu/lunar/missions/orbiter/. [Online]. Accessed 11 Oct 2015

Lunar and Planetary Institute (2011b) The surveyor program. http://www.lpi.usra.edu/lunar/missions/surveyor/. [Online]. Accessed 11 Oct 2015

Lunar and Planetary Institute (2016) Lunar exploration timeline. http://www.lpi.usra.edu/lunar/missions/

Milgram P, Takemura H, Utsumi A, Kishino F (1994) Augmented reality: a class of displays on the reality-virtuality continuum. In: SPIE telemanipulator and telepresence technologies, vol 2351, pp 282–292

Munzee (2012) Munzee: 21st century scavenger hunt. https://www.munzee.com/. [Online] Accessed 20 Oct 2015

NASA (2007) Hiten—Hagoromo mission profile. http://solarsystem.nasa.gov/missions/hiten/indepth. [Online] Accessed 21 May 2015

NASA (2010) Data processing levels. http://science.nasa.gov/earth-science/ earth-science-data/data-processing-levels-for-eosdis-data-products/. [Online] Accessed 21 May 2015

NASA (2011a) GRAIL—NASA. http://www.nasa.gov/mission pages/grail/main/#.Vq9sWvEvtug. [Online] Accessed 11 Oct 2015

NASA (2011b) GRAIL mission overview—NASA. http://www.nasa.gov/mission pages/grail/overview/#. Vq9rw Evtug. [Online] Accessed 11 Oct 2015

NASA (2011c) NASA's recommendations to space-faring entities: how to protect and preserve the Historic and scientific value of U.S. Government Lunar Artifacts. Tech. rep. National Aeronautics and Space Administration, USA

NASA (2013a) ARTEMIS: studying the Moon's interaction with the Sun—NASA. http://www.nasa.gov/mission pages/artemis/#.Vq9qJfEvtug. [Online]. Accessed 11 Oct 2015

NASA (2013b) LADEE—lunar atmosphere dust environment explorer—NASA. https://www.nasa.gov/mission pages/ladee/main/index.html. [Online] Accessed 11 Oct 2015

NASA (2013c) Missions—LADEE—NASA science. http://science.nasa.gov/missions/ ladee/. [Online] Accessed 11 Oct 2015

NASA (2015) Lunar reconnaissance orbiter. http://www.nasa.gov/mission pages/ LRO/main/index.html. [Online] Accessed 21 Oct 2015

NASA GSFC (2008) A standardized lunar coordinate system for lunar reconnaissance orbiter. Tech. rep. NASA Goddard Space Flight Center, Greenbelt, MD

Nava M (2012) The art of journey. Bluecanvas, Inc., Los Angeles, CA

NaviCache (2011) Geocaching with NaviCache—NaviCache home page. http://www.navicache.com/. [Online]. Accessed 20 Oct 2015

Pieters C, Goswami J, Clark R, Annadurail M (2009) Character and spatial distribution of OH/H_2O on the surface of the Moon seen by M3 on Chandrayaan-1. Science 326:568–572. http://science.sciencemag.org/content/early/2009/09/24/science.1178658

Schmitt HH (2005) Return to the Moon. Praxis Publishing Ltd, New York, NY

Snyder JP (1987) Map projections used by the U.S. Geological Survey. Tech. rep. Bulletin 1532, U.S. Geological Survey

Space.com (2007) Space tourism: the latest news, features and photos. http://www.space.com/topics/space-tourism

SPACE.com (2011) Moon packed with precious titanium, NASA probe finds. http:// www.space.com/13247-moon-map-lunar-titanium.html. [Online] Accessed 21 Oct 2015

Spencer S (2012) New game in town. http://www.telegram.com/apps/pbcs.dll/article?AID=/20121110/NEWS/111109916/1246. [Online] Accessed 20 Oct 2015

Spudis PD (2008) NASA lunar exploration: past and future. http://www.nasa.gov/ 50th/50th magazine/lunarExploration.html

Steele JM, Stephenson FR (1998) Astronomical evidence for the accuracy of clocks in PreJesuit China. J Hist Astron 29(1):35–48

Stephenson FR (1997) Historical eclipses and Earth's rotation. Cambridge University Press, New York, NY

Studio NSV (2000) NASA scientific visualization studio. https://svs.gsfc.nasa.gov/. [Online] Accessed 21 May 2015

Sutherland IE (1986) A head-mounted three dimensional display. In: Proceedings of AFIPS 68, pp 757–764

Sven (1996) Prospects of space tourism. In: 9th European aerospace congress—visions and limits of long-term aerospace developments. Aerospace Institute

TerraCaching (2004) TerraCaching. http://www.terracaching.com/. [Online] Accessed 20 Oct 2015

That Game Company (2012a) Journey. URL http://thatgamecompany.com/games/journey/. [Online] Accessed 1 May 2015

That Game Company (2012b) Journey PS3 Games PlayStation. https://www.playstation. com/en-us/games/journey-ps3/. [Online] Accessed 1 May 2015

Tran, T., Rosiek, M., Beyer, R.A., Mattson, S., Howington-Kraus, E., Robinson, M.S., Archinal, B., Edmundson, K., Harbour, D., Anderson, E., the LROC Science Team (2010) Generating digital terrain models using LROC NAC images. In: ASPRS/CaGIS 2010, November 15–19, Orlando, Florida

Wilhelms D (1987) The geologic history of the Moon. Tech. rep. 1348, United States Geological Survey. http://ser.sese.asu.edu/GHM/

Williams DR (2005a) Lunar prospector information. http://nssdc.gsfc.nasa.gov/ planetary/lunarprosp.html. [Online] Accessed 21 Oct 2015

Williams DR (2005b) Ranger to the Moon (1961–1965). http://nssdc.gsfc.nasa.gov/ planetary/lunar/ranger.html. [Online] Accessed 21 Oct 2015

Williams DR (2011) NASA—Clementine project information. http://nssdc.gsfc.nasa. gov/planetary/clementine.html. [Online] Accessed 21 Oct 2015

Williams DR (2013) Lunar exploration timeline. http://nssdc.gsfc.nasa.gov/ planetary/lunar/lunartimeline.html

Williams M (2016) How long would it take to travel to the nearest star?. http://www.universetoday.com/15403/how-long-would-it-take-to-travel-to-the-nearest-star/. [Online]. Accessed 29 Jan 2016

XPrize Foundation (2007) Google lunar XPRIZE. http://lunar.xprize.org/

Zhai P (1998) Get real: a philosophical adventure in virtual reality. Rowman & Littlefield Publishers, Lanham, MD

Zhentao X, Yau KKC, Stephenson FR (1989) Astronomical records on the Shang Dynasty oracle bones. Archaeoastronomy 20(14):61–72

Zooniverse (2013) Zooniverse—real science online. https://www.zooniverse.org/

Ludography

The ludography lists geogames as well as video and online games that are mentioned in the chapters of the book. In addition to games, we also included platforms and frameworks for creating games.

For commercial games, we provide the website of the game as the primary reference. All web links of the ludography have been accessed on June 5, 2017. Games created by researchers are referred to by the earliest publication describing them or by the publication that provides the most comprehensive description. For some of these games, the best description is found in one of the book chapters. In those cases, the reference is simply "Geogames and Geoplay." The ludography distinguishes four types of entries: *C* for console games, massively multiplayer online games or virtual worlds, *P* for platforms used to create and/or run games, *L* for location-based geogames, and *D* for desktop geogames.

Type	Game or technology	Primary reference	Additional reference	Chap. (page)
P	3DGamelab	http://rezzly.com/		10 (6–10)
P	Actionbound	https://en.actionbound.com/		3 (7)
C	Active Worlds	https://www.activeworlds.com		9 (4)
P	ARIS	https://fielddaylab.org	Holden, C (2015) ARIS: Augmented reality for interactive storytelling. In: C. Holden et al. (eds) Mobile media learning, pp. 67–83.	8 (2)
D	B3-Design Your Marketplace!	Poplin, A. (2014), Digital serious game for urban planning: B3—Design your Marketplace!, Environment and Planning B: Planning and Design, 41 (3), pp. 493–511.		4 (11)
L	Can You See Me Now	Benford, S. et al. (2003) Coping with uncertainty in a location-based game. IEEE Pervasive Computing, 2(3), pp. 34–41.		7 (19)
L	CityPoker	Geogames and Geoplay, Chap. 6	Geogames and Geoplay, Chap. 7	3 (8), 6 (2, 6–19), 7 (26–28)
P	CityPokerGD	Geogames and Geoplay, Chap. 6		6 (13–19)
C	Civilization	https://civilization.com/		9 (2)
P	Darkstar	https://github.com/dworkin/reddwarf	Project Darkstar (n.d.). In Wikipedia. https://en.wikipedia.org/wiki/Proje ct_Darkstar#RedDwarf	2 (3)
L	Digital Graffiti Gallery	Holden, C (2015) ARIS: Augmented reality for interactive storytelling. In: C. Holden et al. (eds) Mobile media learning, pp. 67–83.		8 (11)
L	Dow Day	Mathews, J., Squire, K. (2009) Augmented reality gaming and game design as a new literacy practice. In: K. Tyner (ed) Media Literacy: New Agendas in Communication, pp. 209–232.		8 (11)

Type	Game or technology	Primary reference	Additional reference	Chap. (page)
L	Environmental Detectives	Klopfer E, Squire, K. (2004). Getting your socks wet: Augmented reality environmental science. In: Proc. Int. Conf. Learning Sciences, p.614		8 (2)
C	Eve online	https://www.eveonline.com/		9 (4)
L	Feeding Yoshi	Bell, M. et el. (2006). Interweaving Mobile Games with Everyday Life, In: Proc. CHI-06, pp. 417–426.		3 (8)
L	Foursquare (after 2014: Swarm)	https://foursquare.com/ http://swarmapp.com/	Noulas, A. et al. (2011). An Empirical Study of Geographic User Activity Patterns in Foursquare. In: Proc. Int. AAAI Conf. Web and Social Media, pp. 570–573.	1 (15–16) 5 (3, 7–11)
L	Geocaching	http://www.geocaching.org http://www.opencaching.us/ http://www.opencaching.de/	O'Hara, K. (2008). Understanding geocaching practices and motivations. In: Proc. CHI-08, ACM, pp. 1177–1186.	6 (1, 2) 11 (4)
L	Geograph Britain and Ireland	http://www.geograph.org.uk/	Dykes, J. et al. (2008). Exploring volunteered geographic information to describe place. In: Proc. of the GIS Research UK Conf. (pp. 256–267).	1 (16) 5 (3, 7–11)
L	GeoTicTacToe	Schlieder, C., Kiefer, P., & Matyas, S. (2006). Geogames: Designing location-based games from classic board games. IEEE Intelligent Systems, 21(5), pp. 40–46.	http://www.geogames-team.org	3 (8) 6 (13)
P	GIS-MOG	Geogames and Geoplay, Chap. 2		2 (1–13)
D	Green revolution	Geogames and Geoplay, Chap. 2		2 (3, 8)

Type	Game or technology	Primary reference	Additional reference	Chap. (page)
L	Ingress	https://www.ingress.com/	Hodson, H. (2012). Google's Ingress game is a gold mine for augmented reality. New Scientist, vol. 216, no. 2893, p. 19.	1 (8) 3 (7) 5 (3, 7–11) 6 (1, 3) 7 (5) 11 (5)
L	Jewish Jump Time	Rosenkrantz, H. (2014) Jewish time travel gets real. The Covenant Foundation. http://www.covenantfn.org/news/ 152/Jewish-Time-Travel-Gets-Real		8 (11)
C	Journey	http://thatgamecompany.com/games/journey/		11 (12)
L	Mad City Mystery	Squire, K., Jan, M. (2007). Mad City Mystery. Journal of Science Education and Technology, 16(1), pp. 5–29.		8 (2)
L	MapAttack	n.a.	Case, A. (2013), Introducing MapAttack: An Urban Geofencing Game, blog from Oct 17, 2013, http://pdx.esri.com/blog/introduci ng-mapattack/	3 (8)
L	Mentira	Holden, C., Sykes, J. (2012) Mentira: Prototyping language-based locative gameplay. In: S. Dikkers et al. (eds.) Mobile media learning, pp. 113–129.		8 (9)
L	Munzee	https://www.munzee.com	Munzee (n.d.). In Wikipedia. https://en.wikipedia.org/wiki/Mun zee	11 (5)
L	Mystery Trip	Martin, J. (2009) Mystery trip. In: S. Dikkers et al. (eds.) Mobile media learning, pp. 99–110.		8 (9)
L	Neocartographer	Feulner, B., Kremer, D. (2014), Using Geogames to foster spatial thinking, In: R. Vogler et al. (eds). GI-Forum-14: Geospatial Innovation for Society, VDE: Berlin, pp. 344-347.	http://www.geogames-team.org	1 (16) 3 (8) 5 (3, 7–13)
D	NextCity	Poplin, A. (2012). Playful public participation in urban planning, Computers, Environment and Urban Systems, 36(3). pp. 195–206.		4 (7)

Type	Game or technology	Primary reference	Additional reference	Chap. (page)
D	Origami	Geogames and Geoplay, Chap. 3		1 (15) 2 (8) 3 (1–23)
L	Pac-Manhattan	http://www.pacmanhattan.com/	Lantz, F. (2007). Pacmanhattan, In: F. von Borries et al. (eds.) Space, Time, Play, pp. 262–263.	2 (8)
L	Parallel Kingdom	http://www.parallelkingdom.com	Parallel Kingdom (n.d.). In Wikipedia, https://en.wikipedia.org/wiki/Paral lel_Kingdom	7 (5)
L	Pokémon GO	http://www.pokemongo.com/en-us/	Colley, A. et al. (2017). The geography of Pokémon GO, In: Proc. CHI-17, ACM, pp. 1179–1192.	1 (1) 6 (1)
L	Re-activism	Macklin, C., Guster, T. (2012) Re:activism: Serendipity in the streets. In: S. Dikkers et al. (eds.) Mobile media learning, pp. 151–169		8 (14)
L	Riverside	Geogames and Geoplay, Chap. 8		8 (12)
C	Second Life	http://secondlife.com/		9 (4)
C	SimCity	http://www.simcity.com/	Gaber, J. (2007). Simulating planning: SimCity as a pedagogical tool. Journal of Planning Education and Research, 27(2), pp. 113–121.	2 (13) 9 (2)
L	SustainableU	https://mobile.wisc.edu/mliprojects/project-sustainable-u/		8 (11)
P	TaleBlazer	http://taleblazer.org/	http://education.mit.edu/portfolio_page/taleblazer/	8 (2)
D	The Moon Exploration	Geogames and Geoplay, Chap. 11		11 (1–17)
C	The Sims	https://www.thesims.com		7 (3)

Type	Game or technology	Primary reference	Additional reference	Chap. (page)
C	Tomb Raider	https://www.tombraider.com		9 (5)
L	To Pave or not to Pave	Mathews, J. (2010). Using a studio-based pedagogy to engage students in the design of mobile-based media. English Teaching, 9(1), pp. 87–102.		8 (11)
L	Up River	Wagler, M., Mathews, J. (2012). Up river: place, ethnography, and design in the St. Louis river estuary. In: S. Dikkers et al. (eds.) Mobile media learning, pp. 39–60.		8 (12, 15)
L	WeBird	https://mobile.wisc.edu/mliprojects/field-research-project-webird/	https://mobile.wisc.edu/mliprojects/field-research-project-webird/	8 (11)
C	World of Warcraft	https://worldofwarcraft.com		4 (19)
D	YouPlaceIt!	Geogames and Geoplay, Chap. 4		1 (15), 4 (1–22)
L	Zombies Run!	https://zombiesrungame.com/	Zombies, Run! (n.d.). In Wikipedia, https://en.wikipedia.org/wiki/Zombies,_Run!	7 (5, 19)

Printed in the United States
By Bookmasters